Building The PRO STOCK Late Model Sportsman

By Steve Smith

ISBN #0-936834-57-9

Produced and printed in USA

Published By

P.O. Box 11631 / Santa Ana, CA 92711 / (714) 639-7681

www.ssapubl.com

Table Of Contents

Introduction

Race car handling theories and practices change constantly. Racers are constantly researching, experimenting and progressing. New parts come into the marketplace that change the way we do things. Technological changes make us reshape our thinking.

And so it is with the books we publish to help the racer understand his race car. A lot of things in racing have changed since I wrote my first book 18 years ago. With the technological changes come changes in how we think about things. Technology and advancement have changed some of my opinions about what should be done to a car, about how a car should be set up, about the ways we approach some things in handling and suspension theory. That's progress.

Some readers who have a background from reading some of my other books will notice some of these changes of opinion. It's not meant to be a contradiction. It's meant to pass on new knowledge and progress to the racer. Our goal is to make life easier for the racer . . . to help him understand his race car so that racing can be an enjoyable, instead of a frustrating, experience. We want to help the racer put it all together for the optimum level of performance. Hopefully this book will fulfill that goal for you.

Happy racing!

Steve Smith

A Very Special Thanks

No book of this size and magnitude could every be written without the help and cooperation of many special people. I would like to extend a very special thanks to George Gillespie and Doug Day of PRO Shocks, Gary Sigman of Professional Racer's Emporium and his entire staff who so warmly opened the PRE shops to me, Jay Hedgecock of Hedgecock Racing Enterprises, chassis builder and driver John Soares, Warren Gilliland of JFZ Engineered Products, and Dick Anderson of Carrera Shocks. Thank you! I sincerely appreciate the time you took to help me.

Most importantly, a special thank you goes to my very understanding and dedicated family, who have put up with this all-encompassing project for more than a year. Thanks for your caring and support.

Disclaimer Notice

Chapter 1

Rules

The major problem for us in writing this book was to find a uniform set of rules written for the Pro Stock/Late Model Sportsman car all across the country. Not only did the rules vary slightly, but the class is called something different in different locales. In California, it is the Pro Stock class (which we think sounds the most professional and is the most fitting). In other parts of the country, it is a sportsman, late model sportsman, semi-late, cadet, limited sportsman, etc.

To make the contents of this book useable and to fit a wide variety of slightly varying rules, we established the following set of rules which pretty well encompasses those rulebooks we surveyed from various tracks all across the country. Only those rules which effect the design and building of the car and chassis are presented here. The cars we designed and present in following chapters will conform to the prototype set of rules below and to most track rulebooks running this class all over the U.S.

Front Clip

The front frame clip must be a stock type. At least 20 inches of the stock frame forward of the front spindle line must remain stock. The stock frame section must extend back to at least the end of the bellhousing.

Frame

If a stock front clip is used, the rest of the frame may be fabricated from a minimum of 2 x 3 rectangular tubing, 0.090-inch wall section. The fabricated frame section must be of a perimeter type of design with no offset.

Wheelbase

Minimum wheelbase is 108 inches.

Track Width

Maximum track width front and rear is 65 inches.

Minimum Weight

The minimum weight is 3,150 pounds with driver aboard and all fluids full.

Maximum Left Side Weight

Maximum left side weight is 56 percent of total vehicle weight.

Tires

Tires are a racing type with a track management-designated tire to be used by all competitors.

Wheels

Only steels wheels allowed, with a 10-inch maximum width.

Front Suspension

The front suspension must be original equipment type for the type of front clip used. Lower A-arm mounting points may not be relocated. Upper A-arm mounting must be near stock location but may use fabricated mounting plate. Stock type lower A-arms are required. Upper A-arms may be fabricated from tubular steel. Larger ball joints may be used. Steel weight jackers may be added. Stock type front coil springs must be used in stock position. Coils must be at least 5-inch diameter. No coil-overs permitted.

Spindles

Spindles must be stock type only. No fabricated racing spindles allowed.

Steering

Only stock type steering is permitted. No rack and pinion. If original frame clip uses front steering, front steering must be used on race car. If it uses rear steering, rear steering must be used on race car. Stock type of power steering is permitted. Heavy duty components are permitted and recommended. Spherical rod end bearings used in steering or front suspension must be a minimum of 5/8-inch.

Sway Bars

Sway bars may be used. No aftermarket sway bars allowed. Only stock type passenger car or truck bars may be used.

Brakes

Disc brakes are allowed. Only stock type OEM passenger car brakes may be used. If aftermarket or aluminum calipers are used a 100-pound weight penalty must be added to the car.

Shock Absorbers

Racing shock absorbers are permitted.

Rear End

No quick changes permitted. Only a stock type of rear end permitted, such as the Ford 9-inch. Full floating axles are required. No aluminum center sections. Full locker or spool is allowed.

Rear Suspension

Rear suspension may use leaf or coil springs. Coil springs must be minimum of 5 inches diameter. Three-point, truck arm and torque arm suspensions are permitted. Axle damper shocks are permitted. No coil-overs permitted. No spring rod trailing arms allowed. Coil-over units permitted on end of torque arm only.

Roll Cage

Basic four-point roll cage structure in center bay of car must use a minimum of 1.75 inches O.D. steel tubing with a minimum of 0.090-inch wall thickness. All other structural tubing in car affecting safety must be a minimum of 1.5 inches O.D., with 0.090-inch wall thickness.

Engine Mounting Position

Engine and drivetrain must be in center of frame. Engine must be mounted with a minimum of 12 inches from center of crankshaft to ground. Number one spark plug must be within 1-inch of top ball joint center line.

Critical Body Clearance Dimensions

Roof height (minimum height of car from ground) is 47 inches. Minimum 5-inch rocker panel clearance on each side of car. Minimum 4-inch clearance to ground under front spoiler. Maximum of 43 inches from center of front hub to front tip of car.

Chapter 2

Chassis Design

Designing A Chassis

When designing a chassis, the first thing you have to do is decide on the basic specifications and design parameters of the chassis and car. You must first establish the:

1) Wheelbase
2) Tire sizes
3) Minimum ground clearance
4) Front track width
5) Rear track width
6) Roll center height — front
7) Roll center height — rear
8) Center of gravity height (target)
9) Camber change curve
10) Control arm pivot point heights from ground
11) Total weight
12) Weight percentage — left
13) Weight percentage — rear
14) Spindle drop
15) Tire diameter
16) Roll axis line
17) Mass axis line
18) Bump steer

In this chapter we will work through all of these topics and discuss how they affect the chassis design and how they interreact. We will discuss their significance to chassis design, and how each is incorporated into a total design package for a good handling chassis. And, by discussing all of these design topics, we will demonstrate how you can properly design your own Pro Stock chassis that will be a winner. Each topic will be discussed in a point-by-point manner.

Front Suspension Layout

There are a myriad of topics to consider when designing the front suspension geometry layout — spindle choice and its dimensions, spindle drop, kingpin inclination, tire selection (running height, width), wheel offset, frame height and clearance (which influences lower pivot point heights), car track width, camber change curve, static roll center height, and roll axis location.

The major areas of concern in front suspension design are bump steer, camber gain and roll center. The most important parameter to establish first of all is the roll center height and lateral location. The roll center is established by fixed pivot points and angles of the A-arms. These pivot points and angles also establish the camber gain and bump steer.

Defining The Front Chassis

Before we go any further, we must establish the front subframe section that we will be working with. For this class of car, the overwhelming choice is the front subframe from the 1970 to 1981 Camaro.

Why choose this frame section? 1) It is readily available at a cheap price in wrecking yards. 2) Chevrolet parts are widely available and very adaptable to a racing application at a low cost. 3) There are more aftermarket add-on parts available to fit this chassis than any other. 4) There are more heavy duty parts available in wrecking yards to fit this frame than any other. 5) With modification of the upper A-arms and their mounting points, it is easy to adapt this frame section to a racing environment. 6) It's structural layout and weight are very adaptable for a high performance design. 7) The frame width and the curvature of its frame rails make easy adaptation to a race car chassis.

Our Chassis Design

In the following chapter is a blueprint for the building of a chassis for a typical Pro Stock chassis. Beyond major handling considerations, the major design concept for it was the ease of building the chassis by a typical racer in his garage. You will see illustrated in this book the Camaro front stub cut and fabricated to a tubular chassis in a variety of ways by a number of chassis builders and home fabricators. Why did we design ours as we did?

1) Ease of fabrication. Some of the professionally built chassis you will see illustrated in this book employ compound angles in the fabricated tubes which connect the front stub to the frame rails, and in the frame rails themselves. Our design goal was to achieve the same dimensions

and chassis strength employed in these chassis without the need to use sophisticated fabricating techniques, compound cutting angles for chassis tubes, and the use of a chassis jig.

A chassis jig (as shown above) is a must if you are going to use compound angles in your chassis tubing (as below). Our chassis design eliminates this need.

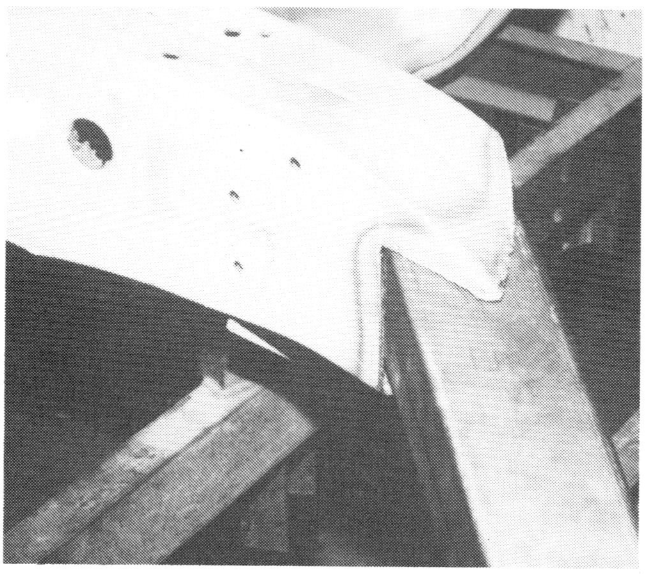

2) We adhered to the rule that at least 23 inches of the stock frame section must remain in front of the front spindles.

3) Our chassis design incorporated the rule that the frame section must remain stock back to at least the front edge of the bellhousing.

4) We considered the objective of ease of repair of the vehicle after any type of crash — easy to cut and replace, easy to reattach, easy to realign any portion.

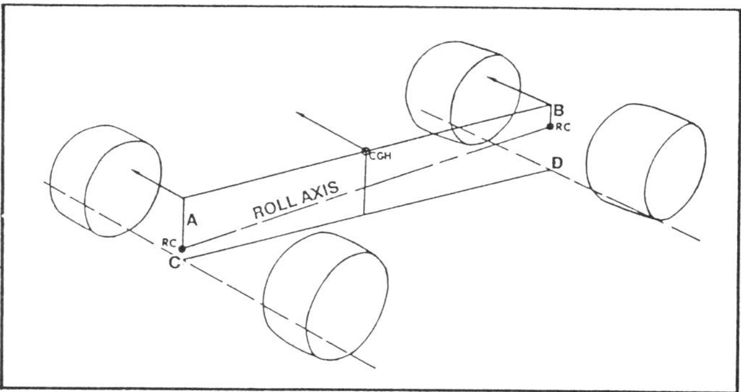

The relationship between the roll centers, roll axis and CGH. A is the front roll center lever arm. B is the rear roll center lever arm. Line CD is the ground plane.

There is a lever arm that exists between the front roll center and the center of gravity height. Assuming a fixed height for the CGH, when the front roll center is raised or lowered, that lever arm (or roll moment arm) increases or decreases. The larger the roll moment arm, the greater the body roll during cornering. The moment arm length does not affect the amount of weight transferred during cornering. It affects the car's resistance to body roll. Body roll must be controlled by spring rates. The larger the moment arm (lower roll center), the stiffer the spring rates must be to resist body roll. The shorter the moment arm (higher roll

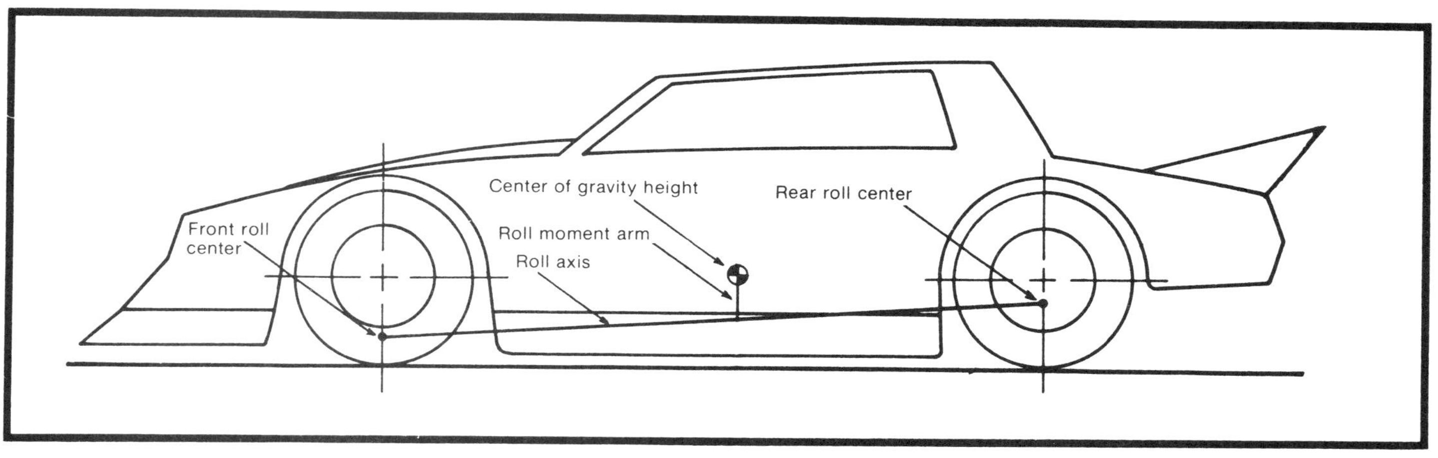

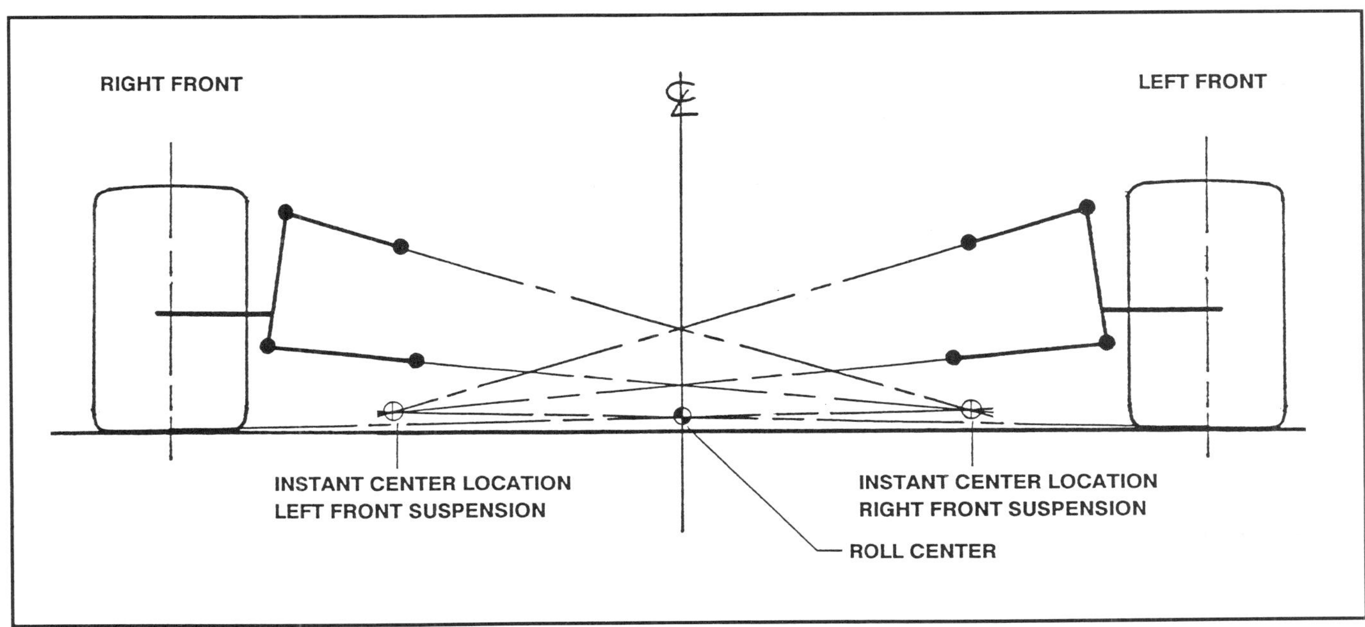

Above, the lay out for the ideal front suspension where the roll center lies on the vehicle center line. Below is an example of how the roll center can be offset from the center line.

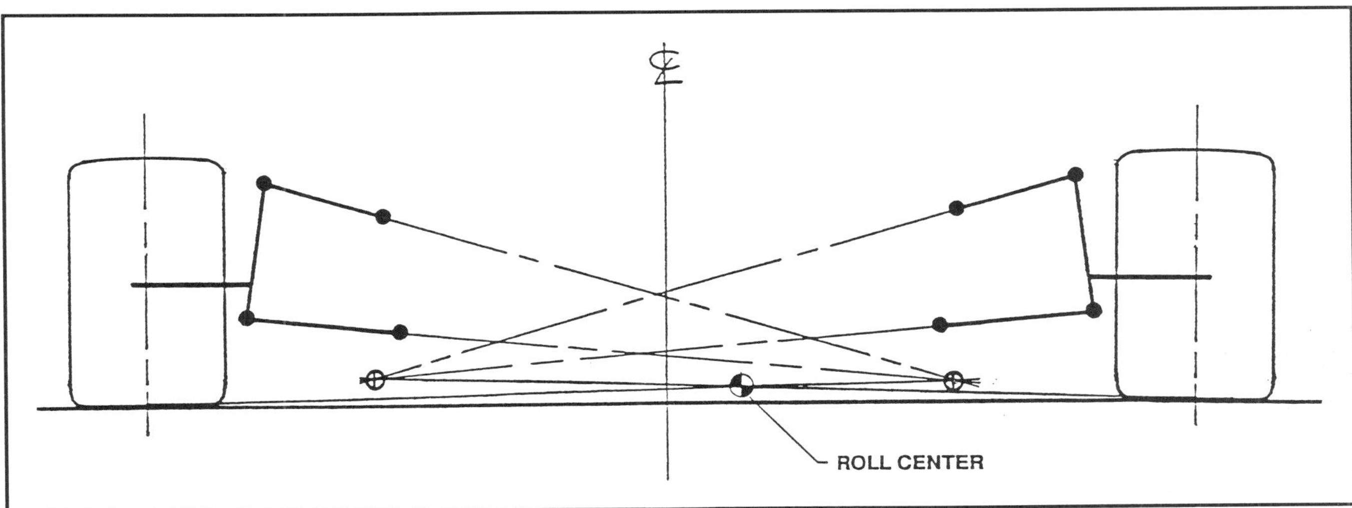

center), the softer the spring rates can be because there is less body roll to resist.

Front Roll Centers

The two most important design considerations in the front end of your race car is roll center placement and camber change curve. Roll center placement includes both vertical and horizontal locations.

Yes, we said horizontal. The roll center does not automatically fall on the physical centerline of the vehicle. It can be displaced to one side of the centerline or the other depending on design considerations. The true roll center of the front independent suspension is that point where the instant center swing arms of the left front and right front cross each other.

You have to consider the left front and right front roll centers together to get a true picture of what is happening in the front end — even if you have a race car that is only turning left. This is because the left front corner is moving as well as the right front during body roll and cornering.

In sketching out the true front roll center of some front ends, it becomes obvious that the roll center can be offset from the physical centerline of the vehicle (see drawing). The offset of the roll center from centerline can actually cause a leverage effect on the chassis (which is undesirable).

Correct Roll Center Heights

The correct design today for front roll centers is between 4 and 5 inches above ground for pavement cars and between 4 and 4.5 inches for dirt cars. These specifications are for cars running on flat to moderately banked tracks. High banked tracks have different requirements (they require a lower roll center so the camber change curve is less).

It used to be thought that dirt track and paved track roll centers had to be very different. Not today.

Asphalt Tracks

The old thinking was to go real high with the front roll center, which allowed very soft spring rates to control the roll rate (body roll) of the car. And thus the ride rate (up and down springing at the wheels) was very soft. But this caused a premature slowing of the performance of the tires. The tires would perform good for 10 to 15 laps, then they went away, slowing up maybe .2 to .4-second per lap due to heating up of the tread face and tire cords. Asphalt track tires are very sensitive. Until tires get up to a certain operating temperature, they get faster. But once they reach a certain point, they start to go away. When spring rates were softer, the tires went away quicker.

The answer to this problem in the last couple of years has been to lower the front roll center and stiffen the front spring rates. The effect of this is that tires have proven to stay competitive longer.

Higher Banked Tracks – Asphalt

When you run a higher banked asphalt track, the front roll center has to be lower — in the area of 2.5 to 3 inches above ground. This is because the cornering forces acting on the car are not creating body roll as much as they are downforce. The main problem here is not controlling body roll but rather bottoming out. With less body roll you need less negative camber gain. A lower roll center produces less camber gain.

The camber change curve of a front suspension is directly tied to the roll center height. The LOWER the front roll center, the LESS camber change per inch of wheel travel. The HIGHER the front roll center, the MORE camber change per inch of wheel travel.

Dirt Tracks

Until recent times, it was thought that higher front roll centers with lower spring rates was the competitive answer for dirt cars. Now, things have changed. Roll centers are back down — between 4 to 4.5 inches — and spring rates are stiffer.

What Influences Front Roll Center Heights?

It was said above that front roll centers have a variance of about an inch — they range from 4 to 5 inches for paved track cars, 4 to 4.5 inches for dirt cars. But what influences how it should fall in that range?

Mass placement in the front end has a big influence. Less front mass requires a lower front roll center. A larger higher front mass requires a higher front roll center. Comparing a Pro Stock class of car to a lighter weight chassis such as an ASA or All Pro car, the Pro Stock car needs a higher front roll center. The reason is that the Pro Stock car uses a heavier engine (cast iron heads, wet sump, steel bellhousing) which creates a bigger mass up high in the chassis.

On the other hand, many Pro Stock cars run with a track tire rule that uses a very hard tire. A hard tire compound requires a lower roll center to create more downforce and bite on the tire. So for our asphalt track blueprint car, we chose the lower range roll center height to use with a harder tire.

Driving style has an affect on your choice of roll center height. A driver with a more erratic style of driving will need — or be more comfortable with — a higher roll center because he can get away with more erratic movements. A low roll center car with the same driver will be more reactive because the erratic loadings and unloadings of the chassis will be amplified with the lower roll center.

A very low front roll center will yield a minimal track width change during body roll and thus minimal lateral tire tread displacement. However low roll centers produce too much body roll, and not enough camber change curve, so we have to come to a compromise.

In light of all this input, what did we choose for our blueprint car, and why? For the paved track chassis, we have used a 4-inch front roll center. The major decision is that this height suits the car for the broadest range of track

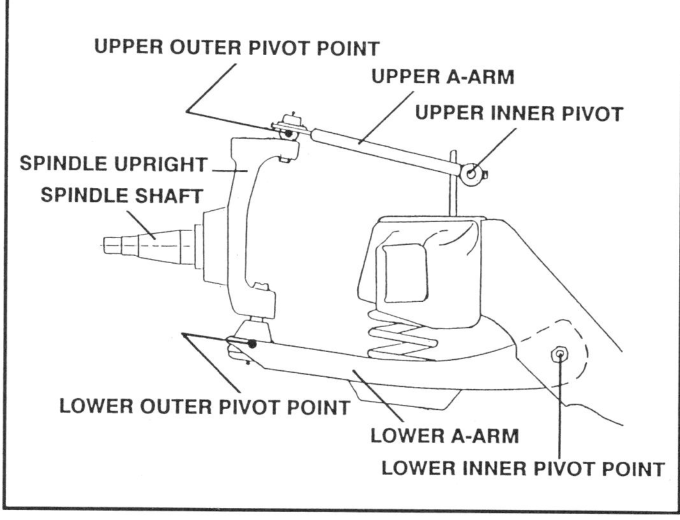

applications. It also worked out to a camber change design that proved to be very workable with the two different spindles chosen for the widest range of track configurations.

For the dirt track chassis, we also used 4 inches for the same reasons as described above for the asphalt track car.

Creating Symmetrical Front Roll Centers

Race cars should be as symmetrical as possible, left to right, as far as control arm angles go. But so many racers put a lot of tilt in their cars (left side lower) which make them assymetrical, and puts a lot of angle in the left side upper control arm. This puts too much camber gain in the left side and not enough in the right side.

The car should be fairly symmetrical as far as the angles of the front A-arms are concerned. A car is more predictable, and faster in traffic, when it is more symmetrical.

Why Low Front Roll Centers Are Detrimental

A lower front roll center (say 1.5 to 2.75 inches) means less negative camber gain and it increases body roll because it reduces the roll resistance. Both of these cases require stiffer spring rates to control body roll. And, the lower the roll center, the more initial static negative camber setting that is required to keep the tire footprint flat in the turns. The flatter the track banking angle, the more the required negative camber setting.

If you run the roll center up higher (4 to 4.75 inches), you don't need nearly as much initial static negative camber. And you don't need as stiff a spring rate at the front corners.

Having less initial negative camber helps a car in the transition turning in to a corner. If a car doesn't have a lot of static negative camber when it enters the turn, it will have a lot more tire footprint on the track at this critical time. If you start with a lot of negative camber, you have less tire footprint on the ground. This will cause an initial understeer during the transition turning in to a corner because there is less of the total tire footprint on the ground as it begins the initial turn in.

The higher front roll center gives the car more negative camber gain without the need for high amounts of static negative camber gain.

Camber gain and camber settings should be designed so that the face of the tire remains absolutely flat on the track at the maximum deflection of the tire.

The higher front roll center (4 to 5 inch range as compared to 2 to 4 inches) is also an advantage with a rough track, because the higher roll center allows the use of softer spring rates, which are much more advantageous over rough surfaces.

How To Design The Ideal Front Roll Center

This section is a step-by-step description of how to draw out in scale, on paper, the roll center location you desire for your particular car and how to relate it to building your chassis. The numbers shown will correspond to the adjacent drawings found on the following page.

You will be making a scale drawing of the front suspension layout of your car. The smallest practical scale you should consider for your drawing is 1/4" = 1" (quarter scale). A larger scale, such as half scale, produces more accuracy, but requires a larger layout table and more specialized drafting tools than most people have at their disposal.

1) Draw the ground line (A on drawing). Establish the desired vehicle track width (our car is 64.5 inches). Mark the track width by drawing the center line of the left front and right front tires (B on the drawing). Draw the mechanical center line of the car (C on drawing). Mark on the mechanical centerline the height above ground of the actual roll center desired (D on drawing). (All numbers we use here are full scale. If you are drawing a car in quarter scale, divide these numbers by 4.)

2) Draw a line from the center of the right front tire contact patch to the roll center height mark on the centerline, and continue to project it toward the left front side. This is the instant center swing arm (E on drawing). We will do only the right front suspension first, then follow the same procedure for the left front.

In the case of the Pro Stock Camaro, the lower inner and outer pivot points are pretty well established (the stock lower arm in the stock location must be used). The lower inner pivot point is the connection of the A-arm to the frame. The lower outer pivot point is the pivot center of the lower ball joint. The only variables are spindle shaft height on the spindle upright, and frame clearance height.

The spindle shaft height (along with your defined tire) will establish the outer pivot point height above ground. The tire diameter will move the spindle shaft center up or down. The frame height, with the fixed pivot point anchor location on the frame, will establish the inner pivot point height above ground. Lower A-arm angle should be designed in the chassis to be flat to slightly uphill (outer pivot point higher than the inner, but no more than one inch higher). This is to minimize the amount of angular change at the bottom of the spindle as the lower outer pivot point arcs upward. With these two points known, a line can be drawn through them to where it intersects the instant center swing arm. This establishes the instant center.

To establish the lower A-arm pivot points on paper, first determine the location of the lower inner pivot point. For our blueprint chassis, this is 7.75 inches above ground, and 8.5 inches outside of the center line of the chassis (the

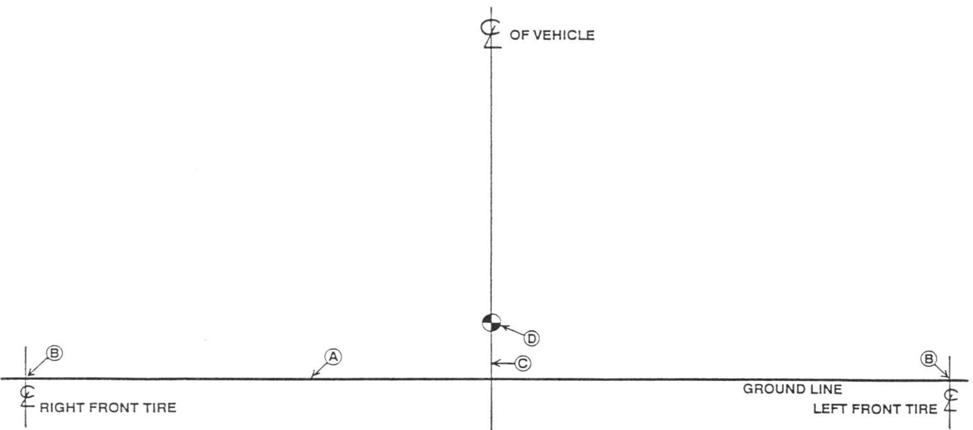

Drawing 1

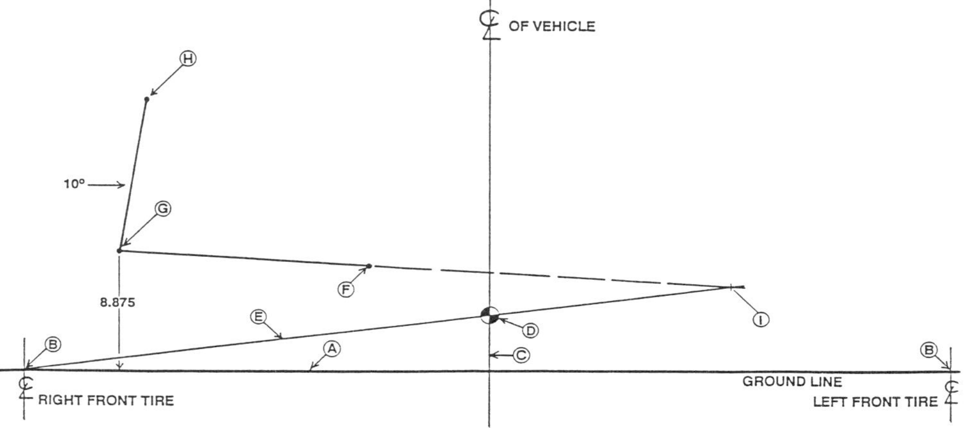

Drawing 2

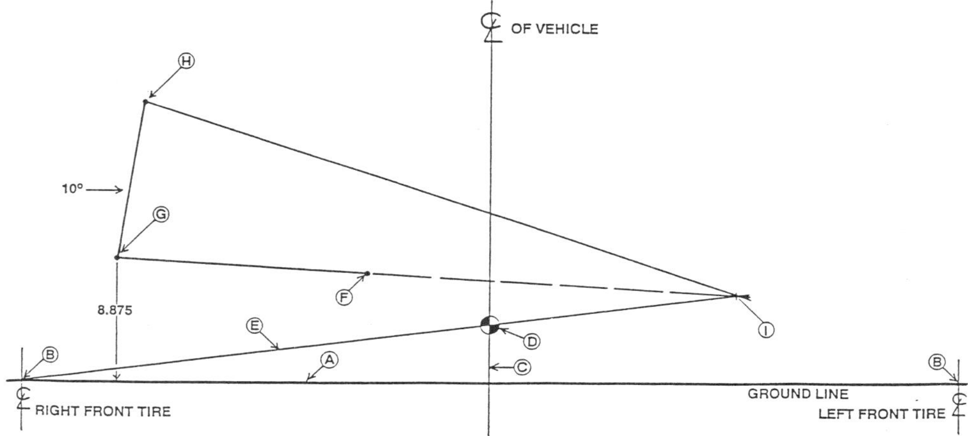

Drawing 3

Camaro chassis is 17 inches from center of left pivot to center of right pivot point). This is marked F on drawing.

The stock lower A-arm is 17.25 inches long, so draw a line 17.25 inches from the center of the inner lower pivot point. This will be the horizontal location of the lower outer pivot point.

To determine the height of the lower outer above ground, first establish the height of the spindle shaft center above ground level. To do this, measure from the center of a wheel mounted with the tire you will be using (remember to allow for proper inflation and compresssion of the tire under load). For our blueprint asphalt car, we used 13.50 inches. From that point, subtract the distance from the spindle shaft centerline to the bottom of the spindle (3.125 inches on the Howe spindle), and the distance from the bottom of the spindle to the center of the lower ball joint pivot (1.125 inches with the Moog K6024 ball joint). Subtracting those two numbers, the height of the lower outer pivot point is 8.875 inches above ground. The two coordinates for the lower outer are now established (G on drawing).

The upper outer pivot point is the next to determined. It is established by adding the overall height of the spindle (8.875 inches for the Howe spindle) to the distance between the top of the spindle and the upper ball joint pivot point (1.375 inches with the Moog K6141 ball joint on the Howe spindle) and the distance between the bottom of the spindle and the lower ball joint pivot (1.125 inches). The total is 11.375 inches.

The 11.375-inch measurement is not made straight up, however, from the lower outer pivot point. The steering axis angle of the

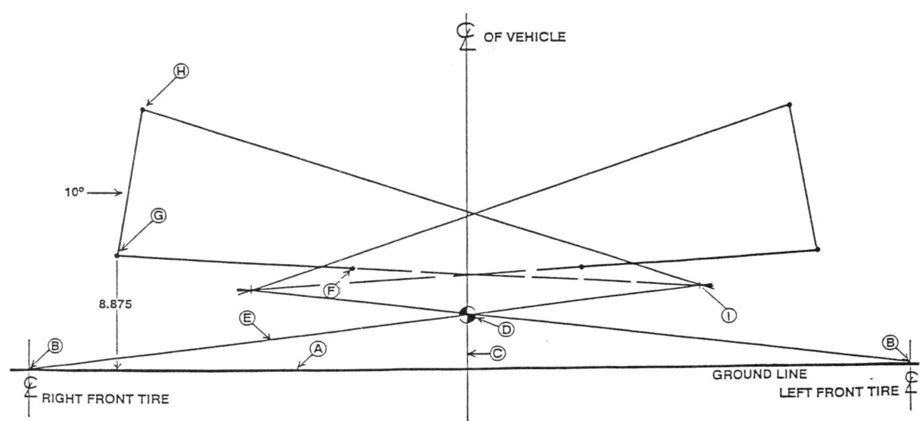

Drawing 4

Measuring the correct pivot point location is critical. The correct location is the CENTER of the ball of the ball joint.

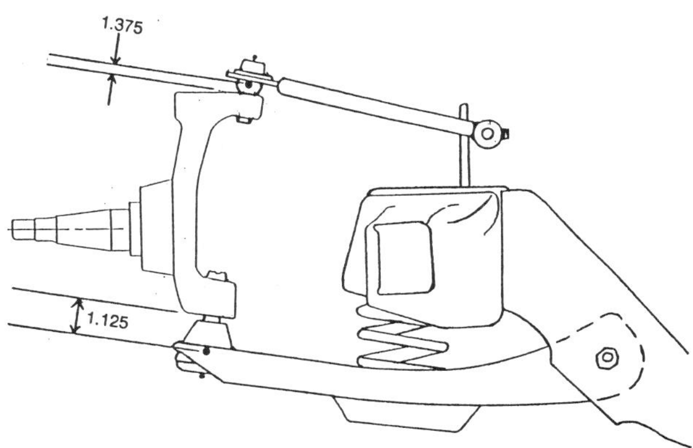

The correct pivot points of the spindle must be located for accurate layout of the front suspension. The pivot points are the centers of the balls in the ball joint housing. Take a close look at the ball joints to accurately locate them. The drawing shows dimensions from the center of the ball to the end of the spindle for a Moog K6141 upper ball joint and a Moog K6024 lower ball joint mounted on a Howe spindle.

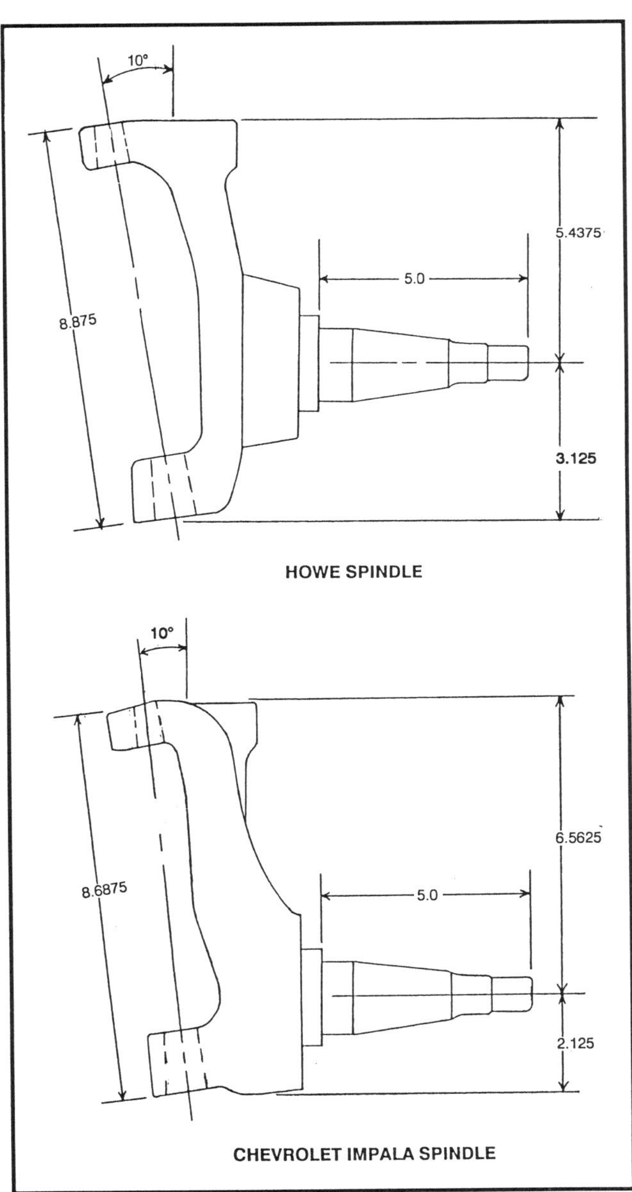

spindle has to be taken into consideration. That number with the Howe spindle is 10°. So draw a line at a 10-degree angle from the lower pivot point, and 11.375 inches up on this angled line will be the upper outer pivot point (H on drawing).

Draw a line from point G to point F on your drawing, and continue to project it until it crosses the instant center swing arm line (E). This is the instant center location (I on drawing).

3) Two lines establish the instant center point — the projected line through the lower A-arm pivot points, and the projected line through the upper A-arm pivot points. The upper outer pivot point (H) is known. Project a line from the upper outer pivot point to the instant center. The

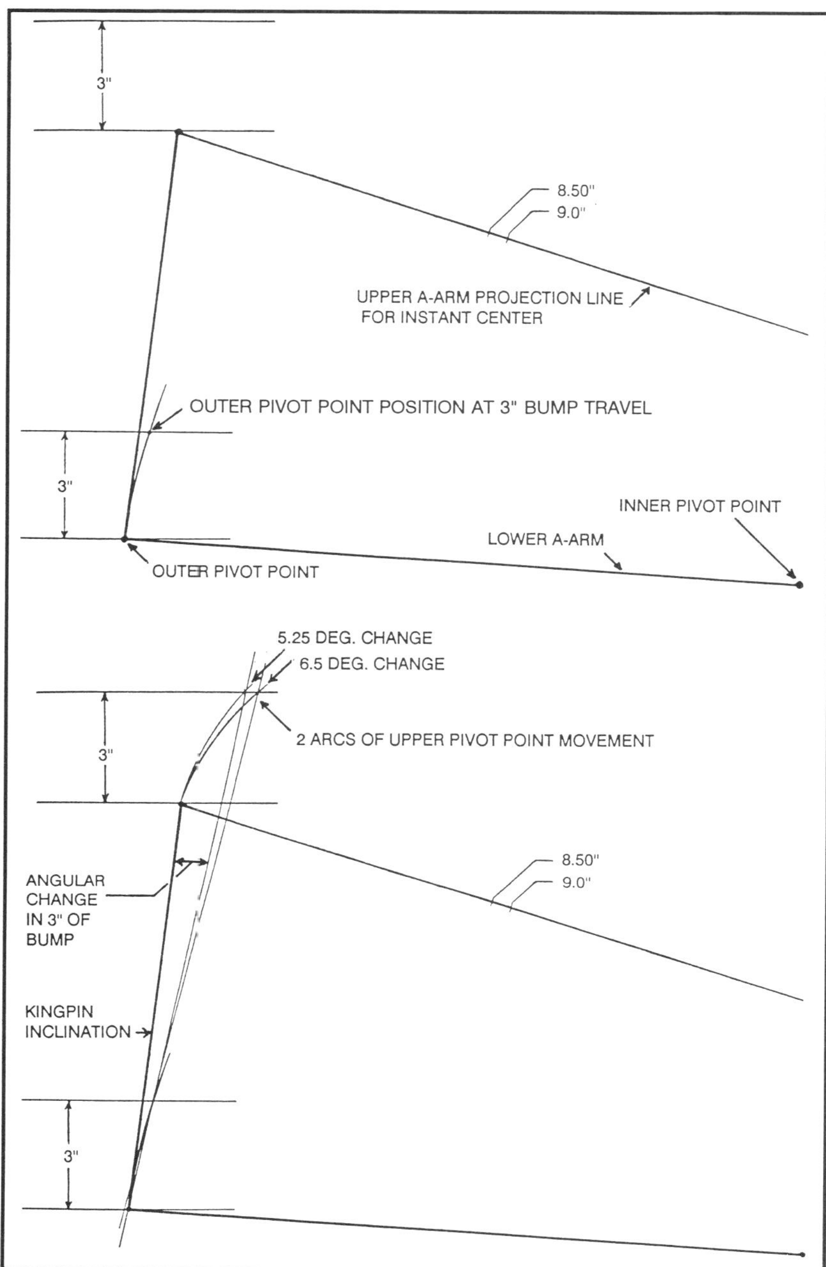

This is how the proper length of the upper A-arm and the camber change curve was determined. A scale drawing was made of the right front corner suspension layout (we used 1/2-scale to insure accuracy). The spindle is then moved on paper through three inches of bump travel by swinging an arc about the inner pivot points of the A-arms. This is where the upper A-arm length is determined. What is desired is 4 to 5-1/4 degrees of angular change of the spindle (negative camber gain) in 3 inches of bump travel (see lower drawing). The inner pivot point location of the upper A-arm is determined by swinging arcs about different locations on the upper A-arm instant center line until the correct angular change can be achieved. The starting point for the upper A-arm length is a calculated guess. In this case, we started with an 8.5-inch upper arm length because that is a very common length on these type of cars. The inner pivot point for 8.5 inches was marked in scale on the drawing and an arc was swung up through 3 inches of bump travel. Then a line was projected straight down from the upper ball joint (at 3 inches of bump) through the lower ball joint (at 3 inches of bump). Measuring from this projected line to the kingpin inclination line with a protractor, it was found that the 8.5-inch upper A-arm created 6.5 degrees of negative camber change in 3 inches of bump. This is too much. The arc is too sharp. This means we need a longer A-arm length.

Next we tried a 9-inch length, which produced a 5.25 degree change. For our project car, we left it at the upper range of the desired camber change because the car will run on a very flat track. If the car was going to be used on a track with more banking, an upper arm length of 9.5 inches would be required.

If the car were being run a on a steeply banked track, a lower front roll center (about 2-3/4 inches) would be used. This would change the angle of the upper A-arm instant center line, so the suspension would have to be drawn out in scale again to determine an A-arm length that would yield about 4 degrees of negative camber gain.

upper inner pivot point will have to fall on that line some place.

4) Repeat this procedure for the left front suspension, keeping it symmetrical with the right front.

To establish the upper inner pivot point location, you simply work out the required length of the upper A-arm. This is done by working out the desired camber change curve (see camber change drawings).

The Camber Change Curve

The camber change curve is the definition of how much wheel camber changes during wheel bump and rebound travel. For a Pro Stock car running on a flat to medium banked track, the correct front wheel should gain between 1.25 and 1.75 degrees negative camber per inch at the right front. Flatter tracks need more camber gain. A completely flat track would need 1.75 degrees per inch. A 10-degree banked track would need 1.25 degrees per inch. A highly banked track would require a camber gain of 1 to 1.25 degrees per inch of bump travel.

There are several factors that will affect the camber change curve for a particular car:

1) The front roll center height and the amount of body roll the car experiences.

2) The front springs stiffness, which controls the amount of body roll.

3) Sidewall height and sidewall stiffness of the particular tire being used.

4) The amount of lateral displacement on the face of the tire at the contact patch.

5) Wheel deflection.

6) Tire width in relationship to rim width (sidewall cantilever deflection).

More deflection — from any cause — means more negative camber change required.

Tire Construction Influence On Camber Curve

The camber change curve must be designed precisely for the type of tire the car is going to use. Hard economy tires typically have harder sidewalls and stiffer belts in them, and they won't deflect the contact patch nearly as much as one of the "open track" types of tires in which the sidewall wedge is shorter and the compound is softer. A retread tire (which is typical of many Pro Stock classes) falls into the hard economy tire category. The larger and softer (more flexible) "open" tire needs more negative camber gain per inch of bump movement, and they also need more static negative camber (because of instantaneous deflection at the contact patch when they are steered). Both of these are to keep the inside edge of the tire from pulling up off the ground.

As an example, let's say we take a sample car and run it first with economy tires and then with the open type tire. With the economy tire, you stand the tire upright more — less static negative camber. And you use less stagger in the rear (stagger equates to negative camber on a solid axle). Then put the "open" tires on the same car. To get it to perform correctly, you will have to add static negative camber at the right front (it will probably require 1.25 to 1.75 more degrees of negative camber), and it will need more stagger in the rear (it will need at least 3 to 3-1/2 inches rear stagger). Increased tire deflection is reflected at the rear end as increased stagger required. Tire stagger also effects the attitude of how the car comes off a corner. See the Chassis Adjustment chapter for more information on stagger use.

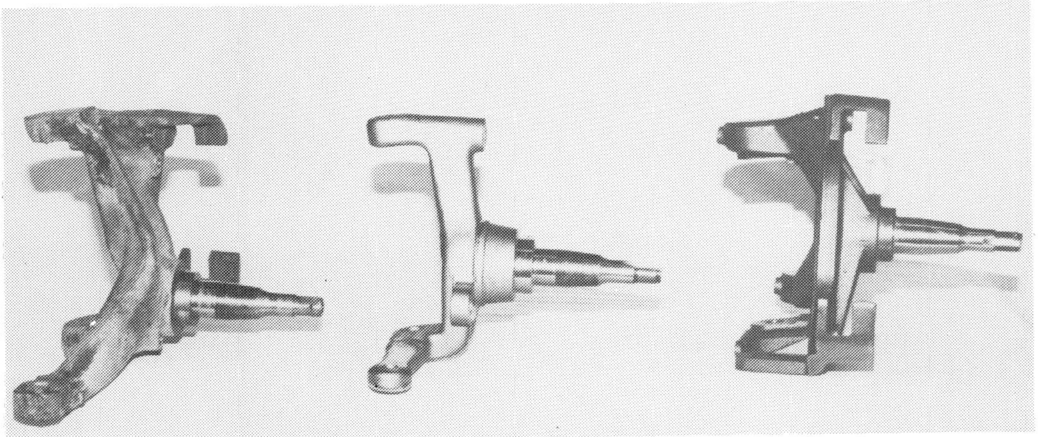

A vivid illustration of spindle drop between three different spindles. At left is a stock 1971-1976 Chevy Impala spindle. In the middle is a Howe spindle. At right is a PRE UltraDrop spindle.

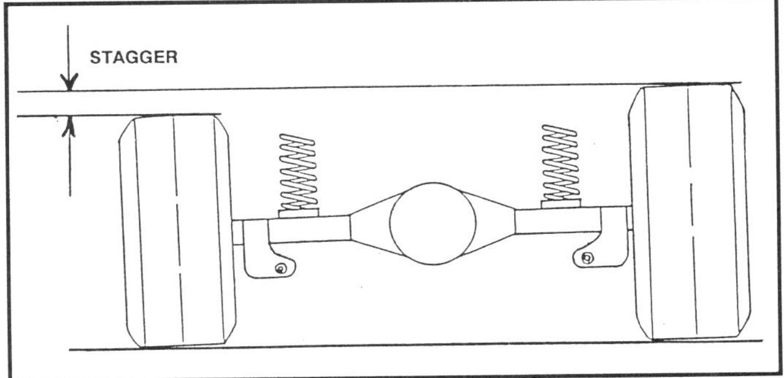

Rear tire stagger creates negative camber for the tires.

Spindle Drop

Spindle drop is the measurement of the spindle shaft centerline from standard position up the spindle upright.

Relocating the spindle shaft upward drops the car down, with all other influences staying the same.

The standard position of the spindle shaft center line for the Chevrolet Impala spindle (a very popular stock spindle choice for this class of car) is 2.125 inches above the bottom of the spindle forging. The Howe forged spindle has the spindle shaft center line located 3.125 inches up from the bottom of the forging, which equates to a 1.0-inch spindle drop from stock location. And PRE offers a spindle (which they call the Ultra Drop) which has the spindle center line located 4.75 inches up, which equates to a 2.625-inch drop from stock location. (See the Notes on Components subchapter in the Chassis Construction chapter for a complete definition of all spindles referenced.)

Spindle drop can be a very helpful modification to a chassis. But, be aware that you cannot just bolt on a pair of dropped spindles and go racing. There are a lot of chassis change implications in using something like this. Take a look at our asphalt chassis blueprint in the next chapter, for instance, where both the Howe and the Chevy Impala spindles have been used. The chassis utilizing the Impala spindles has to have the front stub attached to the chassis differently than the one using the Howe spindles, simply because of the difference in geometry caused by the two different spindles.

When the spindle shaft is relocated on the upright, many changes take place in the front end geometry. All suspension pick-up points are changed, and upper and lower A-arm angles are changed. There are most probably chassis clearance problems created as well. When a dropped spindle is used, the entire front suspension geometry must be designed around it. You must draw everything out in scale on paper and see what the suspension does as it moves in bump (at the right front) and rebound (at the left front).

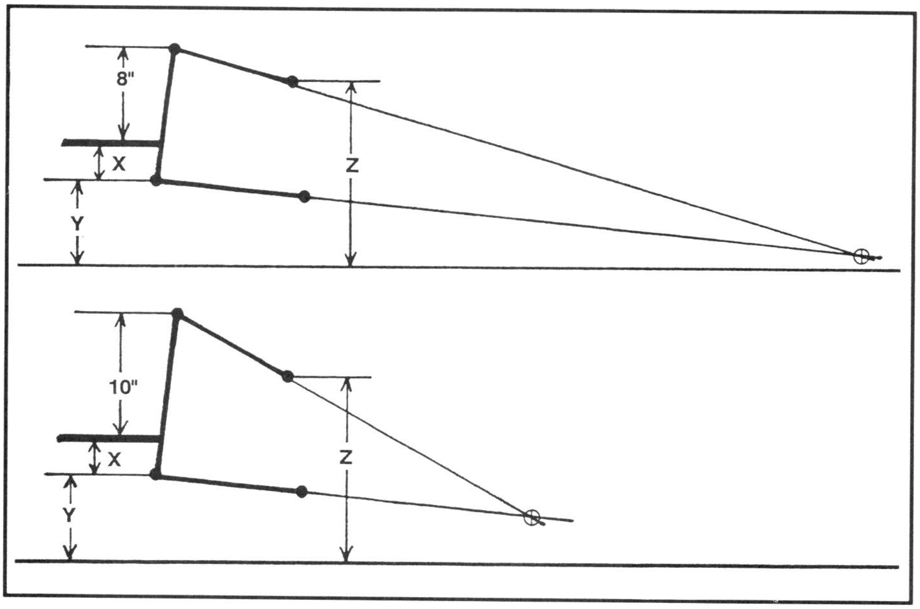

Spindle Overall Height

The overall height of the spindle upright is another design consideration for front end geometry. A shorter spindle upright height (upper ball joint mount to spindle shaft center line is shorter), with all other pivot points and ground clearance remaining the same, will lower the roll center height. This shorter spindle will produce a lower roll center height because the geometry creates a longer instant center. The shorter the instant center to the wheel it is drawn from, the higher the roll center will be. If you are faced with a situation of having a tall spindle, the only way to correct for it is to relocate the upper inner mounting point higher.

Bump Steer

Bump steer is a front suspension geometry condition which results from the steering tie rod moving in a path which is dissimilar to the path in which the wheel it is connected to is moving. Bump steer causes the wheel to be turned when it encounters bump, or upward movement of the wheel, even though the steering wheel is not turned. This results from the wheel and the tie rod end moving through paths, or arcs, which are different.

The accompanying drawings illustrate how the tie rod can be designed and positioned so that bump steer is minimized or controlled.

For short track cars (both asphalt and dirt track applications), bump steer should be 0 at the left front (no steer whatsoever through bump and rebound), and 0.040-inch out per inch of upward spindle travel at the right front.

Why should there be bump out at the right front? The right front of the car is the controlling side for steering. The deeper you drive into a turn, the more bump (upward) travel there is at the right front. And the closer you get to the apex of the turn, the more the car is going to

If points X, Y and Z remain the same and only the height (8 versus 10 inches) changes, the shorter spindle will produce a lower roll center height because the geometry creates a longer instant center. The only way to correct for this is to relocate the upper inner mounting point higher.

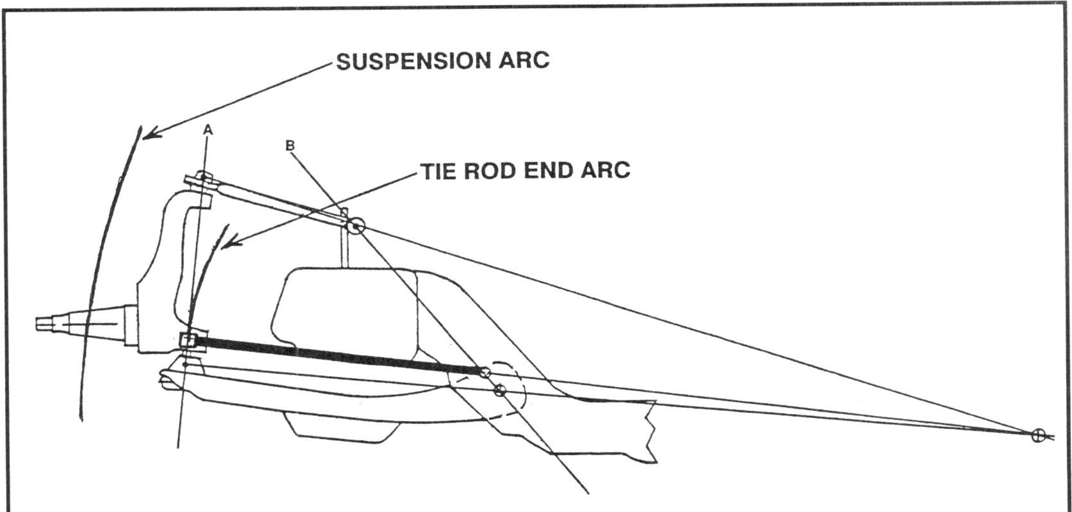

Bump steer is perfect when the suspension arc and tie rod end arc is parallel to each other. To do this, the tie rod length must fall between the confines of planes A and B, and the tie rod's center line must intersect the instant center. The tie rod may be laterally displaced so long as the length remains the same as if it were between A and B planes.

push. The trick is to get the car to turn at the apex without pushing so you can get back on the throttle immediately. The key to getting the car to turn is more toe-out at this point. The more toe-out you have here, the more the car will cut down without pushing. As the right front bumps out, what is really happening is the driver can steer the car to the left more as he keeps the right front in line with the right side of the car. So as the car rolls over it is trying to bump out at the right front, the driver is steering more to the left, and this is creating more toe-out at the left front. The toe-out is really showing at the left front even though it is bumping the right front out. This is a controlled bump out because the more you compress the right front of the car, the more it toes out. This is not static toe-out, which can be detrimental on the straightaway.

This concept is very closely related to Ackerman steering. Be sure to read the section on Ackerman steering in the Front Suspension and Steering chapter.

Measuring and adjusting for bump steer is done after the chassis is built. We have covered that topic in the Front Suspension and Steering chapter.

Rear Roll Center Location

The rear roll center location for a 3,150-pound Pro Stock car should fall between 10 and 12 inches above ground. The ultimate rear height location is dependant on the front roll center location. A dirt track car will have a slightly lower rear roll center height than a pavement track car because it helps create side bite on dirt. We chose to use an 11-inch rear roll center height for the dirt track blueprint car, and 12 inches for the asphalt blueprint car. Our front roll center height on both cars is 4.0 inches.

The rear roll center has its own "separate identity" from the front end. You get the front roll center established first because that one is the most critical, and the most difficult to change. Then work with the rear roll center. Make your best guess for rear roll center height, then do track testing to confirm your choice. Make sure you construct the brackets for the rear lateral locating device so the roll center can be moved up or down at least an inch in each direction.

Rear Roll Center Lateral Location

The roll center of a suspension system is that point at which lateral forces can be applied without creating rolling of the sprung mass. It is, simply, the center of a rolling object. The center will stay still while the outsides will move in equal arcs. In coil-sprung stock cars that have a Panhard bar as a rear lateral locating device, the length and mounting location of the Panhard bar determines the rear roll center. If, for example, the race car has a short Panhard bar which anchors to the chassis on the right and to the rear end housing at the center of the car, the roll center is not at the center of the vehicle but rather halfway between the attaching points of the Panhard bar. In this case, the rear roll center would actually be located about 12 to 15 inches to the right of the mechanical center line of the car. This would create a roll axis for the car which runs at an angle to the center line of the car (see drawing next page).

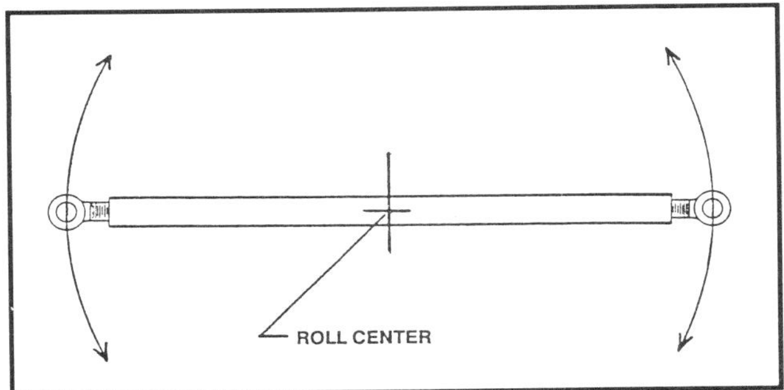

The Panhard bar length determines the lateral location of the rear roll center. The roll center is the absolute center of the Panhard bar.

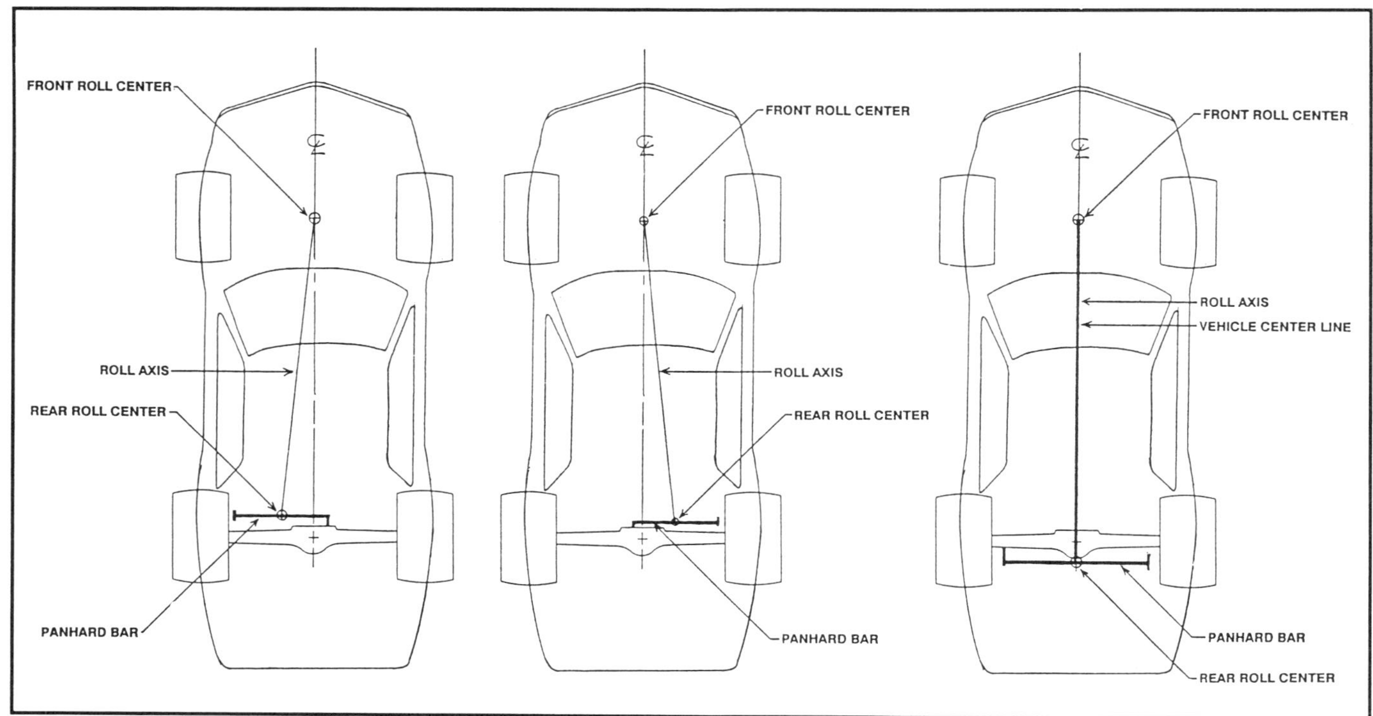

The rear roll center location with a solid rear axle depends not only on the geometry of its layout but also is influenced by relative spring rates. Having a stiffer spring on one side than the other will move the roll center toward the stiffer spring. How much does the unequal spring rate move the roll center? We calculated an example using a 60-inch rear track width, a 40-inch rear spring track (center of the left rear spring to center of the right rear spring), a 200 lb/in spring at the left rear and a 250 lb/inch spring at the right rear. The roll center was shifted to the right (toward the stiffer spring) 2.05 inches.

It is apparent, then, that if a short Panhard bar is used which offsets the roll center to the right, a stiffer left rear spring can be used to offset it back to the center line of the car. Something on the order of at least a 50 lbs/inch stiffer spring at the left rear should be used with the short Panhard bar.

What happens if the short length Panhard bar is attached to the left side of the frame instead of the right side? The roll center will be offset to the left of the vehicle center line, and it will now require a stiffer right rear spring than a left rear. If this were a high cross weight car, this would not be a desirable situation, because the higher left side weight would require a stiffer spring at the left corner.

Some Rules Of Thumb For Rear Roll Centers

1) A highly wedged car (high cross weight, in the area of 58 to 60 percent) along with a short Panhard bar mounted on the right will keep the rear of the car very tight. The left rear spring should be stiffened to compensate for this.

2) Cars with less cross weight, in the range of 52 to 55 percent, perform better with the rear roll center in the center of the car or offset toward the left rear tire.

3) If you use stiff spring rates, you can run lower roll centers. Or, turning that around, if you have lower roll centers, you are going to need stiffer spring rates to control the car.

4) Rear roll center location should be somewhere between the bottom of the ring gear (assuming the use of a Ford 9-inch rear end) and just below the axle housing tube. So the ballpark figure is somewhere between 10 and 11.5 inches. Be aware that tire height is going to influence the roll center height. If you bolt on a set of tires that have a lower section height, you will lower the roll center. A taller tire (such as most dirt track tires) will raise the roll center on the same chassis. It is best that you have a Panhard bar (or other attaching linkage) that is adjustable to handle this.

5) Rear tire width has an influence on rear roll center height, especially on a dirt track. When you have a small rear tire, the rear roll center should be lowered to get more bite. When you have large width rear tires, you want to get the rear roll center up higher, or else the car won't turn at the apex. With a low rear roll center and big tires, you will have too much rear traction, and thus a chassis push. For comparison, with 10-inch wide rear tires, the roll center should be between 10 and 11.5 inches. With an 8-inch wide rear tire, the rear roll center would be between 8 and

9 inches. And with a 13-inch rear tires, the rear roll center would be more in the neighborhood of 12 to 13 inches.

Front Vs Rear Roll Center Height Relationship

The front roll center should always be lower than the rear roll center. If the front roll center is too much lower, the car will roll too much too quickly in relation to the rear upon turn entry (most probably creating a push, depending on other factors). On the other hand, if the front roll center is right and the rear roll center is too high, the car is going to tend toward oversteer upon turn entry (with all other factors being equal). Wheelbase length is also a design consideration in roll center height relationship. The shorter the wheelbase, the more critical the front/rear roll center relationship is.

There is no set mathematical relationship between the front and rear roll center heights. The only way to design any race car is to use sensible ballpark guidelines for the front roll center as outlined before. The rear roll center should also be built within ballpark guidelines (between 10 and 12 inches). Then only actual testing will show you the ultimate placement for the rear roll center. The rear linkage which sets the roll center height should be moved in 1-inch increments, with a test session for each placement. As track testing gives you feedback about how different roll centers affect your car's handling, make notes about this for future fine tuning help. (In the Chassis Adjustment chapter, there is a discussion of using the rear roll center as a fine-tuning device.)

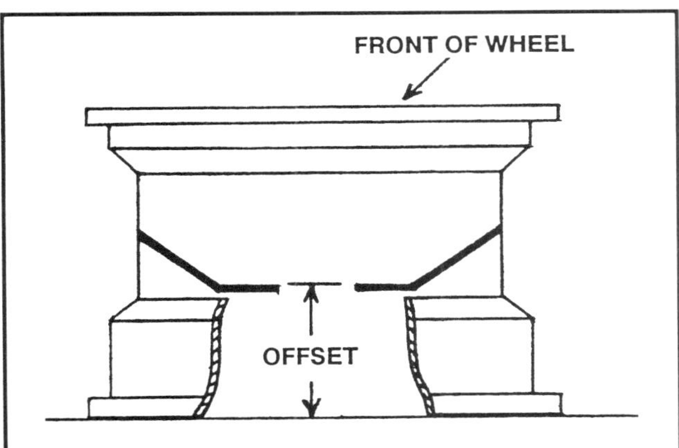

Wheel offset is the back spacing of the wheel. To measure, lay the wheel with the back side down on the shop floor. Measure from the floor to the back of the center section of the wheel. This is the offset. If you are not restricted by an offset rule, use a larger offset on the right side and a narrower offset on the left, such as 5 inches on the right and 3 inches on the left, to gain left weight percentage.

Front Vs. Rear Track Width

For a short track late model sportsman (both dirt and paved track applications), a slightly narrower rear track width creates a more stable handling car. The narrower rear track makes the car tighter both on turn entry and acceleration on turn exit. We are using a 64.5-inch front track width and a 62-inch rear track width on this book's blueprint chassis.

An independent front suspension also is improved by having a wider track because it provides a wider support base for the car.

When running on a track that is real fast and fairly highly banked where you keep up the momentum through the turns, you want to have a rear track width equal to the front. This is so the car doesn't bind up and scrub off speed. The easiest way to handle this is to change the right rear wheel to a different offset that moves the right rear out in line with the right front.

Roll Axis

The roll axis is an imaginary line connecting the front roll center with the rear roll center and is theoretically the axis about which the body rolls during cornering. The ideal chassis design would bring the roll axis to the mechanical center line of the car. This would make the car roll equally left to right about one axis line. To increase or descrease bite, then, you could play with left and right spring rates. There would be no need to add or subtract spring rate at one corner or the other to support extra weight leveraged from a further roll center.

Mass Axis Vs. Roll Axis

The mass axis can be pictured as a line running from front to rear, as seen from above, connecting the center of the major concentrations of weight in the vehicle. It would be like slicing the car into several sections, like a slice of bread, and locating the left to right center of weight in each slice, then connecting all these centers with a line. This axis will not be a straight line, but can be projected into one to give an indication of its relationship to the mechanical car center line and roll axis.

Every car has its own identity and you have to adapt roll centers, mass placement and spring rates to what the car likes. Only testing will tell you for sure. The ideal chassis design would put the roll axis and the mass axis very close to each other.

If you cut a hole in the roof of a car and position a crane so it could hook the actual center of gravity, it would pull the car off the ground perfectly balanced. This is the center of all weight masses.

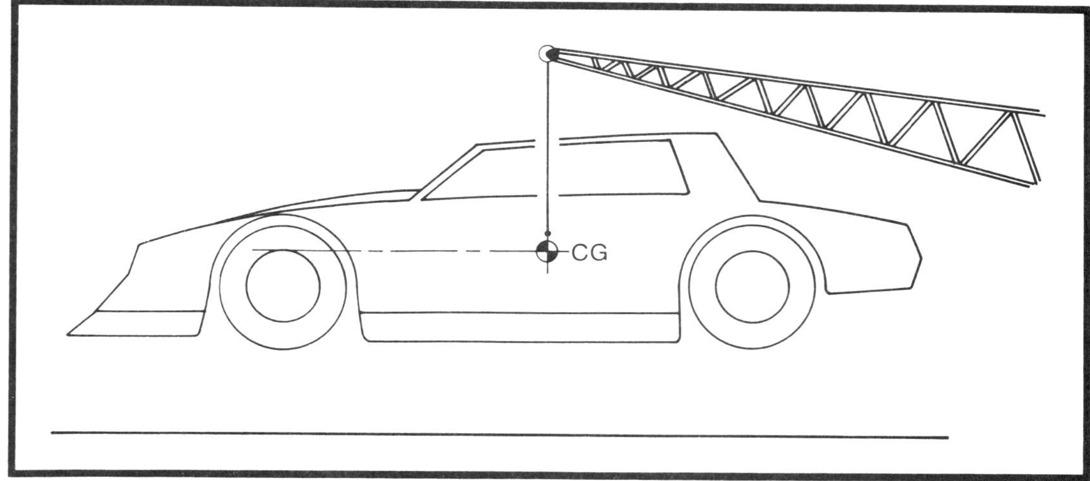

Center Of Gravity Height

The center of gravity of a car is that imaginary point which is the absolute center of all weight in the car — vertical, front to rear, left to right.

The center of gravity height (CGH) is the balance point in the chassis which evenly splits the upper and lower weight masses of the car. We concern ourselves with the center of gravity height because it is that point through which the centrifugal force acts during cornering. The higher the CGH is above the roll axis, the more weight transfer from inside to outside during cornering.

For a car running on a paved track, the CGH should be as low as absolutely possible. For paved track Pro Stock cars, this usually ranges between 15.25 and 17.5 inches, depending on engine height and ballast placement. For dirt track cars, the CGH should be between 16 and 20 inches. The lower CGH figure for dirt is for a dry/slick track, where set-up conditions are very much like those for a paved track. The higher CGH is used for a wet/loose dirt track where downforce and side bite is required on the right rear tire.

For a dirt track that starts out real wet and loose, and ends up dry/slick by the time the main event is run, it is advisable to have movable ballast locations in the car. This way the ballast can be bolted in up high for the wet surface condition, and then can be moved downward as the track starts to dry, ultimately getting the ballast very low in the chassis for the dry/slick condition.

Determining the precise CGH of your car can be very difficult if you do not have access to wheel scales. If you want to closely approximate the CGH, use the following guidelines: for most Pro Stock class cars, the measurement from the center line of the camshaft to the ground is very close to the CGH figure. For very highly modified chassis built with a dry sump system and with the ultimate placement of weight in mind, and with a large amount of ballast placed very low in the chassis, this approximated CGH can be lowered by 2 to 2-1/2 inches.

Polar Moment Of Inertia

A very important handling concept which dictates the willingness of a car to change directional attitude is called

A paved track car's CGH should be as low as is possible. It usually falls between 15.25 and 17.5 inches on a Pro Stock car.

A dirt track car's CGH should fall between 16 and 20 inches. The lower CGH is more advantageous for a dry slick track, while the higher CGH is best for a wet track.

the "polar moment of inertia." "Poles of inertia" is just another way of saying "center of weight concentration." The "moment" in this concept is determined by the front-to-rear (or longitudinal) location of the center of gravity, and how far the masses are located from it. A car with a low polar moment of inertia is one that has fast directional maneuvering response. The closer the concentration of masses is to the CG, the less resistance of the vehicle to changing directions. A high polar moment car would be one with a lot of weight overhanging the wheels at each end of the car. It is very difficult to coax it into changing directions.

How does this concept apply to a Pro Stock car? A low polar moment of inertia is definitely desirable to create smoother maneuverability of the car. Build your race car and chassis light — as light as you possibly can (of course, don't ever sacrifice on safety). Then, if you've done everything right, you can add 350 to 400 pounds of ballast to the finished car.

Where do you put all that ballast? Bolt it to the closest point to the center of the car that makes the car weigh correctly for the left and rear percentages you want.

How high should the ballast be? It depends on the type of track and track conditions:

Dirt — track is wet/loose: It should be high (20 to 25 inches from ground) to help get side bite.

Dirt — track is dry/slick: It should be as low in the chassis as possible. This loosens the car up when it gets hard to turn.

Asphalt track: It should be as low in the chassis as possible.

The Rear Suspension System

Rear Spring Location

For this type of car, we have considered using only stock-type coil springs for the rear suspension. There are too many problems associated with using leaf springs. The major problem is that the leaf spring has to handle too many controlling functions in the rear suspension — functions that can be divided up and controlled separately with coil spring rear suspensions. The leaf spring must handle not only chassis sprung weight support, but also must control rear axle torsional wrap-up and lateral location of the rear end against the chassis. And because of the design and materials of leaf springs, their rate goes away quickly and they must be replaced. Leaf springs also make it difficult to make small changes in chassis height at the rear corners, to jack weight from the rear corners, to control the amount and rate of pinion angle change, and to make spring rate changes in small increments. Using coil springs overcomes all of these problems.

The ideal rear spring location is behind the axle and below the center line of the rear axle.

However, there is a place for leaf springs in Pro Stock cars. They can be used by the novice racer who doesn't desire to make, or is confused by, the track tuning adjustments required with a coil spring type of rear suspension. Leaf springs are more suited to dirt track applications than paved tracks.

Ideal Rear Coil Spring Location

The rear coil springs should be mounted as far out toward the wheels as possible for the best control. Keep the springs mounted low, on a bracket mounted to the rear end housing. They should be mounted, at the highest, at the center line of the axle housing, and preferably lower. Do not mount the springs on top of the rear end housing. This gets the chassis roll points too high for effective weight transfer leverage. Mount the springs behind the rear end housing, and mount the shock absorbers in front of it.

Spring Base To Wheelbase Relationship

The spring base is the distance from the center of the front wheels back to the center of the coil spring. A longer spring base to wheelbase relationship is preferable for any type of coil spring rear suspension. It makes the car slightly less reactive and helps the driver gain a better seat-of-the-pants feel of the car during cornering.

Rear Suspension Attaching Linakges

There are four different movements that the rear end can experience: side-to-side, fore and aft (front to rear), pitch about its axis (rolling forward or rearward), and rotation about a vertical axis (a vertical axis is one up and down

The ideal three-link rear suspension has two bottom links each 24 inches in length and an upper link which is 3/4 the length of the lowers.

through the center section, and the rotation is one which is seen from looking down from above).

Because there are four different directional movements that the rear axle can experience, the maximum number of linkages which can be attached to it is four. If any more linkages than that are used, the axle is subject to mechanical bind as it goes through its movements.

The rear axle restraints for a paved track coil-sprung car include two trailing links (which control fore and aft movement and rotation about a vertical axis), an upper trailing link (which controls pitch about the axle's axis), and a Panhard bar (which controls lateral or side-to-side movement). This is a description of the basic three-link rear suspension system.

3-Link Rear Suspension Layout

The major design considerations with a 3-link system is the lengths of the three links. The two bottom links should be 1.5 times longer than the upper link (or in reverse, the upper link should be 3/4 the length of the lower links).

The bottom links should be 20 to 24 inches in length. If you can accommodate a 24-inch lower link in your chassis, by all means use that length. Longer is better because it produces less angular change on the rear end housing as one side moves in bump and the other moves in rebound.

Some chassis designers have used a longer link on the right side than the left. This is for a roll steer effect. Don't do it. Number one, it simplifies things a great deal when you can carry one spare link and it will fit the left or the right side. Number two, without working out all of the intricacies of the geometry on both sides, you could build in some roll steer problems without knowing it.

The ideal 3-link layout would be 24-inch long lower links running parallel to the ground and a 16-inch upper link running downhill (front end of it lower) at about a 7-degree angle.

You should avoid any lateral angles with the lower links (pointing the links in or out as they run forward). These lower links are what propel the mass of the car forward from the rear tires. They push the race car around the track. If the links run at various angles, it dissipates the energy being used to push the car forward. The lower links should push the car forward without binds.

A 3-link rear suspension can be designed so that is functions very similarly to a torque arm suspension. That is, the upper link can be angled downward severely so that its projected intersection point with the lower links will fall just under the neutral axis line. This will create a very heavy loading of the rear tires in leverage against the chassis. But a 3-link suspension has one big drawback when it is hooked up for maximum tire loading under acceleration — severe wheel hop and lightness under braking. That linkage relationship which really hooks up the rear wheels under acceleration creates the opposite and equal reaction when the loads are transferred the opposite way under braking. The rear end will get very light, causing a lack of

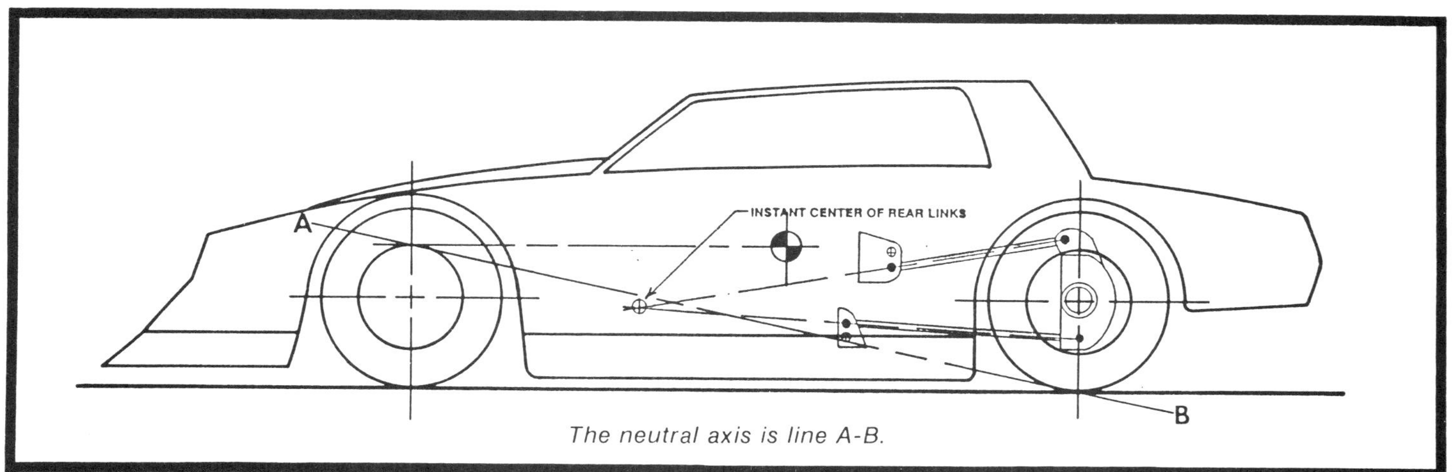

The neutral axis is line A-B.

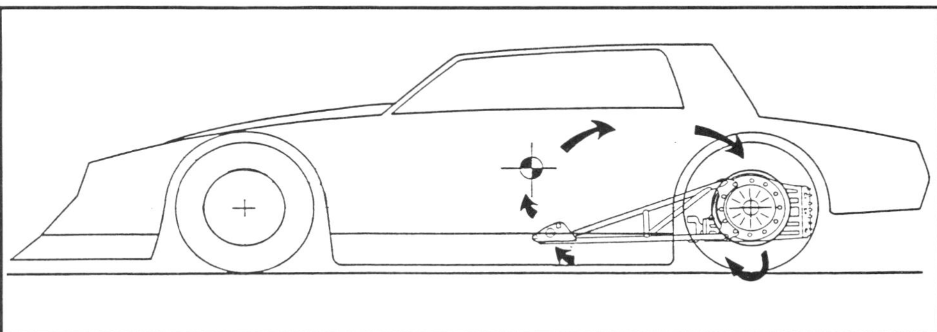

When the rear axle housing is free to rotate about its own axis in birdcage brackets, the power of this rotating housing can be used as leverage against the weight of the chassis to solidly plant the rear tire contact patches.

directional control plus wheel hop under braking. This is the reason that a maximum of 7 degrees downhill angle is used for the upper link.

Dirt Track Rear Suspension

The basic linkage arrangement for dirt track cars starts with the same basic 3-link system as described above for paved track cars. But the addition of a torque arm to this system makes a world of difference for getting a car hooked up on a dirt track. Torque arms, no matter what kind you use, create more rear bite than any other type of system.

The introduction of the birdcage or floating bracket to the rear end allows one direction of axle freedom whose energy can be harnassed for more bite through the use of a rigid torque arm.

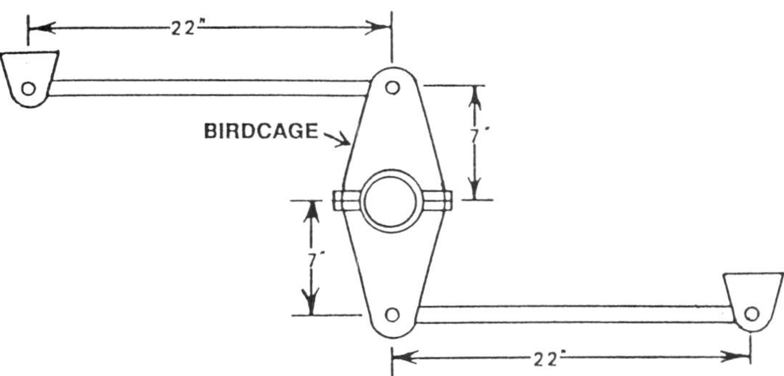

The birdcage bracket is retained by equal length arms which are equally spaced from the center line of the bracket.

Let's look at how the rear axle operates with birdcages or floating brackets. The rear end housing must still be restrained at the left and right side by some type of trailing links to control rotation about the vertical axis and fore and aft movement. Most generally, the birdcages are controlled and restrained by a Watt's linkage type of arrangement, which has one rod running forward and an equal length rod running rearward. This creates an up and down axle movement with no steering angle on the axle. Another popular type of restraining linkage for floater brackets is the wishbone link rod (see more information about it in the Rear Suspension chapter).

The beauty of this system is that the rear axle housing is free to rotate about its own axis within the bearings of the restrained birdcages. The mechanical power of this rotating axle housing, under acceleration, can be harnessed with a torque arm without the interference of any other arms attached to the housing. This torque arm uses leverage against the weight of the chassis to solidly plant the rear tire contact patches.

These three links (two trailing links plus torque arm) combined with a Panhard bar for lateral control make up the four directions of rear axle control required.

Our only caution on torque arm systems is using them on real tight short tracks. They can create so much rear bite that the push condition which is created can be uncurable. This can definitely depend on driver style too. In this case, a 3-link rear suspension system might be more preferable.

Torque Arm Length

Torque arm length depends on rear weight percentage plus wheelbase length. If the torque arm is too short and the rear weight percentage is too small, the rear end of the car will lift in the air when the driver stands on the gas. The opposite side of this is too much rear weight percentage and too long a torque arm. In this case, when the driver stands on the gas, the nose of the car will rise. Neither case gives good performance.

The correct approach is to have a torque arm length that falls about 7 to 8 inches behind the horizontal center of gravity. This horizontal CG can be computed by multiplying the wheelbase by the front weight percentage. The rule of thumb is the larger the rear weight percentage, the shorter the torque arm length required because the horizontal CG is located further behind the front wheels. If you have a 58 percent rear weight, a 36 to 38-inch long torque arm would be in order. If you have a 50 to 53 percent rear weight, a 42-inch long torque arm would be better suited.

Lateral Locating Linkages

The Panhard Bar

The full length Panhard bar is the simplest and most widely used of the lateral control linkage systems. The bar is attached to the chassis at one side of the car and to the rear axle housing at the opposite end.

As we have seen previously in discussing rear roll centers, the Panhard bar establishes the rear roll center. The roll center lies at the center of the bar, no matter how long the bar is or where it is attached to the axle. Because the rear roll center should be located, most ideally, at the car's mechanical center line, a full length Panhard bar should be utilized to properly locate the roll center, and to minimize rear roll steer.

Roll steer in the rear end is introduced with the use of a short Panhard bar. This happens because the arc described by the short bar is sharper than the arc described by the rear end as the car rolls. The other drawback of the short Panhard bar occurs in one-wheel bump along a straightaway. As a bump is encountered and one wheel moves up and down, the difference in arcs between the Panhard bar and the axle causes a lateral displacement of the tires. This means the rear tires are pulled left and right across the track surface. This scuffing will cause excessive tire heat and tire wear. The rougher the track surface, the more this will effect the tire wear and bite.

If for some reason you are forced to have to use a short Panhard bar, mount it with the right side chassis attachment point slightly higher than the inner attachment point on the rear axle. The higher outer pivot point will drop to create a parallel bar as the body rolls and the right side drops. This will help minimize the rear steer problem.

The Watt's Linkage

The Watt's linkage is the only type of lateral control linkage which will result in zero roll steer. It is an ideal type of linkage to use for a paved track Pro Stock car.

The Watt's linkage has a bellcrank which is attached to the center of the rear end housing on a pivot. It has two arms from the bellcrank, one extending in each direction, ultimately anchoring in a bracket on each side of the chassis. Each end of these arms is free to pivot in spherical rod end bearings.

This system keeps the chassis centered in its proper position over the axle during all movements of bump and rebound, without any bump steer resulting from a linkage arm moving the outside wheel through a forced arc.

Using the Watt's linkage properly requires the use of an "over and under" frame structure at the rear of the car. That is a structure where there are frame rails both above and below the rear axle housing. This type of structure gives an advantage for properly mounting the chassis brackets for the linkage arms.

A Watt's linkage operates from a bellcrank mounted solidly on the rear end housing with arms anchoring it to a bracket on either side of the frame.

Using a Watt's linkage requires the use of a space frame or "over and under" frame rail arrangement so brackets can be conveniently located to attach the linkage arms.

Ideal Shock Absorber Mounting Locations

Front Shock Mounting Location

The more travel the piston in the shock absorber has per inch of wheel travel, the more effective the damping will be, regardless of the rate of damping. To achieve this, it is imperative that the front shock be located as close as possible to the lower ball joint. So many chassis builders hang the shock absorber on the side of the front A-arm. In order to gain wheel clearance with this type of mounting, the shock has to be positioned at least 50 percent of the A-arm

When the front shock absorber is positioned on the outside of the lower A-arm, it is too far away from the wheel to be effective. It should be anchored just in back of the lower ball joint and run straight up through the upper A-arm to attach to the chassis.

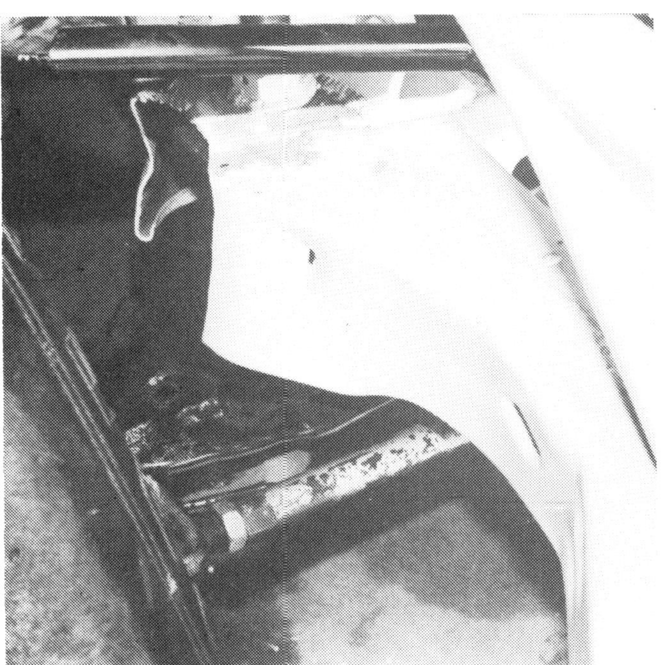

The prohibiting factor for running the shock absorber up through the middle of the upper A-arm is that the outer protrusion of the spring pocket is in the way. It can be trimmed away to gain needed clearance.

The bottom of the rear shock absorbers should be mounted 8 inches below the center line of the rear axle housing, and 4 to 5.5 inches in from the back face of the brake rotor.

distance back to the mounting point of the arm. This makes the shock absorber almost totally ineffective.

The proper place for mounting the front shock absorber with the Camaro front stub is just inside the lower ball joint on the lower A-arms. This is the closest point to the wheel that is possible with this frame. The side of the coil spring pocket on the frame will have to be trimmed (see photos) in order to accommodate the shock in this position. This will not be detrimental to the structural integrity of the frame. With the lower end of the shock absorber mounted as close as possible to the lower ball joint, position the top end so the inward mounting angle falls between 10 and 20 degrees from vertical. 15 degrees would be ideal. With the car setting at normal ride height, the shock absorber should be mounted at one half its total travel. The stoke of the shock should be positioned so the shock never reaches the end of its travel or bottoms out during the full amount of wheel travel.

Rear Shock Mounting Location

The rear shock absorbers should be mounted ideally at 8 inches below the center line of the rear axle housing. This is because it is easier to transfer weight front to rear because the top shock mount location is lower in relation to the CGH of the car, making it easier to compress the shock absorber (by greater leverage ratio between the CGH and the top shock mount height). The bottom mount

Shock absorber mounts are very heavily stressed. This well-designed upper mount has triangulated bracing in 2 planes.

should also be between 4 and 5-1/2 inches away from the back face of the brake rotor.

The shock mounting angle should be 15 degrees from vertical. It should not vary more than a couple of degrees from this figure in either direction.

Both the front and rear shock absorbers should mount either with tie rod ends or spherical bearings. Never use a rubber-mounted shock absorber. The tie rod end shock absorber is generally a less expensive piece. The spherical bearing end shock is more expensive, but is generally available in more damping force rates and is a more quality piece. Let your budget be your guide.

Chassis Structure

Chassis Rigidity

The chassis should be totally rigid so it does not act as one big torsion bar. If a chassis flexes, it is absorbing suspension input movement that should be controlled by the springs and shocks. Chassis flex gives you inconsistant handling and feel. Improved chassis rigidity means suspension adjustments are much more responsive. A stiffer chassis means smaller adjustments to affect the same change. A rigid chassis means adjustments — such as springs, shocks, weight distribution and wedge — will make meaningful changes to the set-up.

Rear Space Frame

The rear frame bay section, in so many chassis designs, doesn't get the support it requires to handle the rear input loads and chassis front-to-rear weight transfer. A typical

The rear space frame, or "over and under" frame rail arrangement produces a much stiffer and stronger rear bay section.

design has a frame rail that angles up and runs straight back over the rear end housing. This is half of a proper space frame section design. The other half would be a frame rail properly tied in that runs under the rear axle housing. This is sometimes called an "over and under" frame rail design. If you can run this type of frame rail design according to your track rules, by all means do it. It produces a much stiffer and stronger rear bay section.

This type of frame section is legal in all of the track rules we investigated for this type of car, so we integrated it into our blueprint design.

Suspension Component Design

Suspension locating links — A-arms, trailing arms, Panhard bar, etc. — must be designed and located so that all loads on them are in straight tension or compression, without any bending moments. These links should be connected to the chassis over as wide a base as possible to spread the loads going into the chassis.

The suspension links, and their attachment points, must not have any compliance. This is a function of stiffness of the link material as well as the attachment points. All attaching links in the Pro Stock class car (tie rods, trailing links, etc.) should be steel and not aluminum as in lighter weight cars.

All chassis components mounted to the frame and its tubes should be mounted in double shear — that is, having mounting bracket material on either side of the attaching component (such as a rod end) with a bolt passing through the entire assembly. This adds considerably to the strength

Double shear, or straddle mounted, brackets have bracket material on both sides of the attaching linkage. This makes the attachment much stronger than a single shear mount as seen below.

of the assembly because the shear force is spread over the area of the two brackets instead of just one.

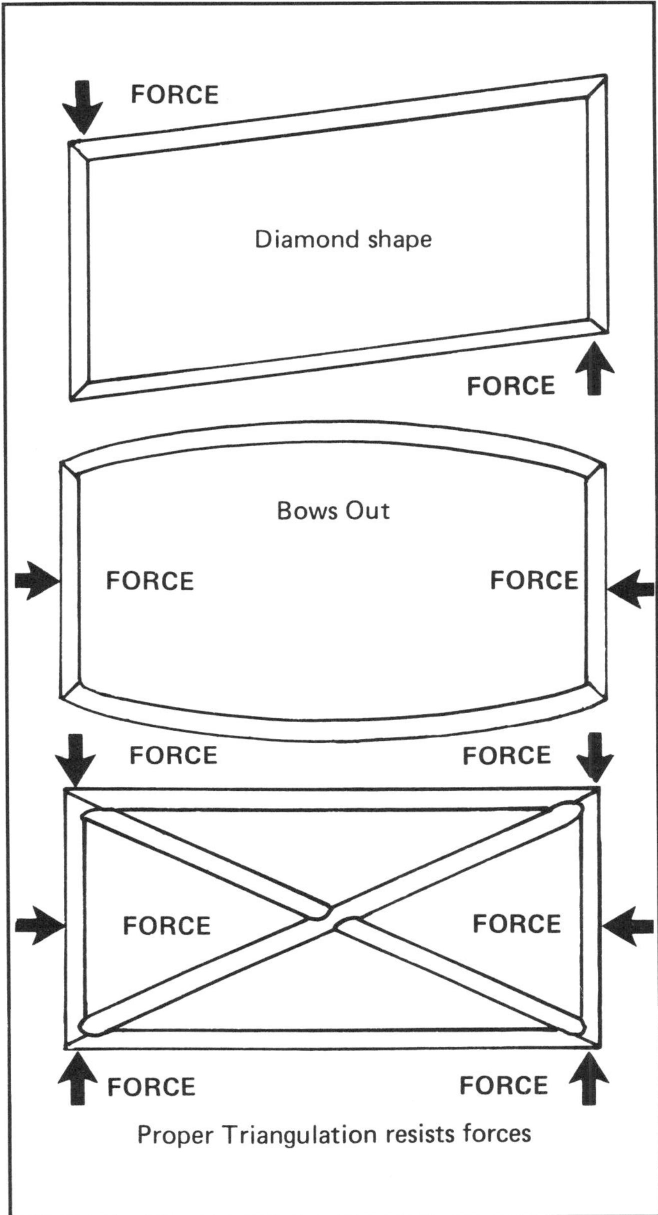

Proper Triangulation resists forces

Cage Structure Rigidity

In order to assure a rigid chassis for a car that will produce predictable handling results, all suspension input points must be well connected to the triangulated cage system. We have observed many chassis designs that are inadequate in this area — most notably in the shock absorber mounts. The best shocks in the world cannot do their intended job adequately if the input loads are just flexing an unsupported chassis tube.

The basic box design of a roll cage is just a rectangular structure (actually three rectangles in a stock car — front bay or engine compartment, middle bay or driver compartment, and rear bay). That retangular structure is very inefficient, in terms of chassis rigidity and reinforcement, until tube triangulation is introduced into it. But don't get carried away with adding too many tubes. Make each tube work for your design in the most efficient way possible, such as resisting bending and twisting loads simultaneously, so that you have the maximum stiffness per pound of structure.

Roll Cage Design

The major criteria of designing a roll cage is to first establish the basic cage as required by the racing association's rules. From this basic framework, tubes can be added for additional safety and to stiffen and strengthen the chassis. Triangulated tubing members should be incorporated which connect all the suspension pick-up points, using tubes structured in V and W shapes to spread the input for-

Notice how the tubes are used efficiently in the chassis above and below to form triangles which resist bending and twisting loads which are input by the suspension.

ces fully throughout the chassis. This prevents the input forces from being concentrated in certain areas, which would create bending and flexing moments in certain main tubes. The V and W shaped arrangements of tubes form structurally stronger designs because they transfer input forces over a much wider area. Think of triangular designs as load spreaders.

The total cage design should be fully triangulated with an absolute minimum of bends to assure total rigidity. A minimum number of bends in the tubes also makes the cage much easier for the average racer to construct in his shop.

The cage and chassis should also be designed to provide ease of maintenance to the complete car and all components. Things like ease of engine removal, spring and shock removal, and trailing arm mount access should be carefully considered. You will have to work on the car and chassis — make sure it is easy and comfortable to do so.

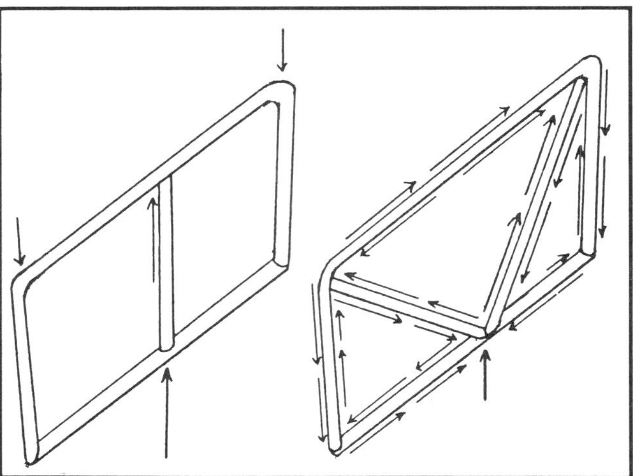

The basic principle of triangulation is to spread the loads evenly throughout the entire structure. At right is shown how a load fed in is evenly distributed and resisted. When the arrows are pointing against each other on the same tube, the load is placing the tube in compression. This is the round tube's strongest asset. In the configuration at left the load places the tubes in bending, which is a round tube's weakest point.

Roll Cage Building Tips

Use web-type gussets to tie the main chassis roll cage tubes together. These improve load distribution over a wider surface area.

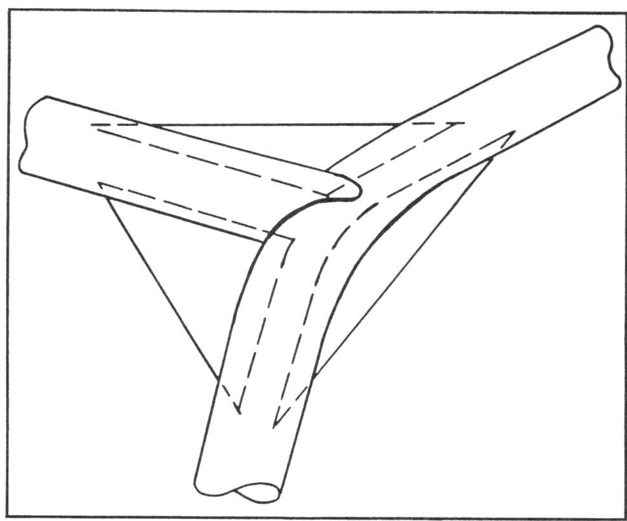

Web-type gussets used to tie the main chassis roll cage tubes together improve the load distribution over a wider suface area, increasing strength and helping to prevent separation under impact.

A diamond shaped flat plate gusset over a butt-welded joint greatly improves the strength of the joint by spreading the loads over a much wider area.

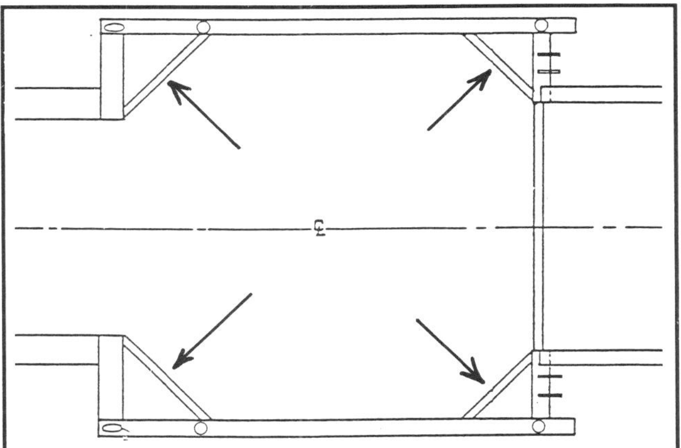

It is important to add diagonal bracing at square cornered joints to prevent "diamonding" or shifting.

Use a diamond-shaped flat plate gusset over a butt-welded joint. This improves the strength of the welded joint and spreads loads over a wider area.

At square tube square-cornered joints, add a small diagonal piece to prevent shifting or "diamonding."

Don't leave tube ends open on a chassis. Weld on a plate-type cover of the same material as the tube. This improves the torsional strength of the tube, and ties it in as a stronger unit to any adjacent tube. And, a closed tube prevents the entry of water which will cause rust (which seriously deteriorates the strength of the tube).

One of the most important bracing tubes in the entire chassis is the leg brace. This is a curved tube that extends from the outer protrusion of the left side door bars forward to the front sub frame. The purpose of it is to protect the driver's feet and legs from the intrusion of the left front wheel if it is pushed back hard in a crash, or from the intrusion of another car that hits behind the left front wheel.

Attaching brackets to the roll cage tubes should aim the input loads at the centerline of the tube. This is so the bracket and the suspension piece input loads are fed into the center of the tube rather than the side of it.

Note the closed cap on the end of the frame rail tube (arrow). Also note that the brace on the right side added before the door bars adds considerable strength to the center bay.

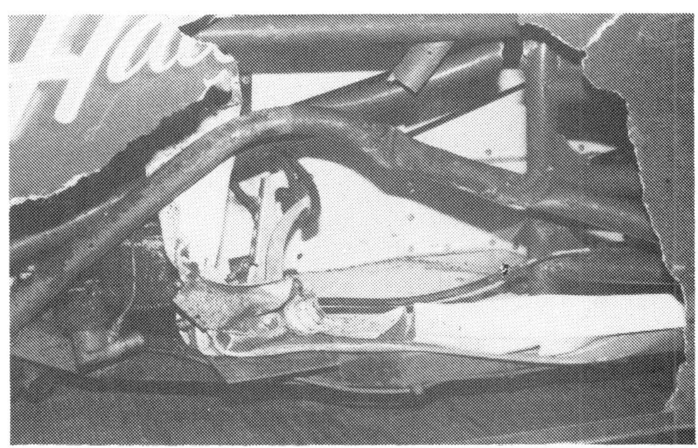

The leg brace tube is very important for the safety of the driver. Can you imagine the injuries the driver of this car could have sustained if that bar hadn't been there?

What's the difference? If the loads are aimed at the center of the tube, they are put in tension or compression only. These are the tubes' strongest attributes. However, if the bracket is placed out to the side of the tube and the suspension input loading is fed up and down on the bracket, it moves the tube in torsion (twisting) which is its weakest point.

When trailing arms connect to a bracket and/or a crossmember which resides very close to the driver, put a deflection cover over the bracket where the trailing arm connects. This protects the driver should the trailing arm or its rod end break and be pushed upward or forward.

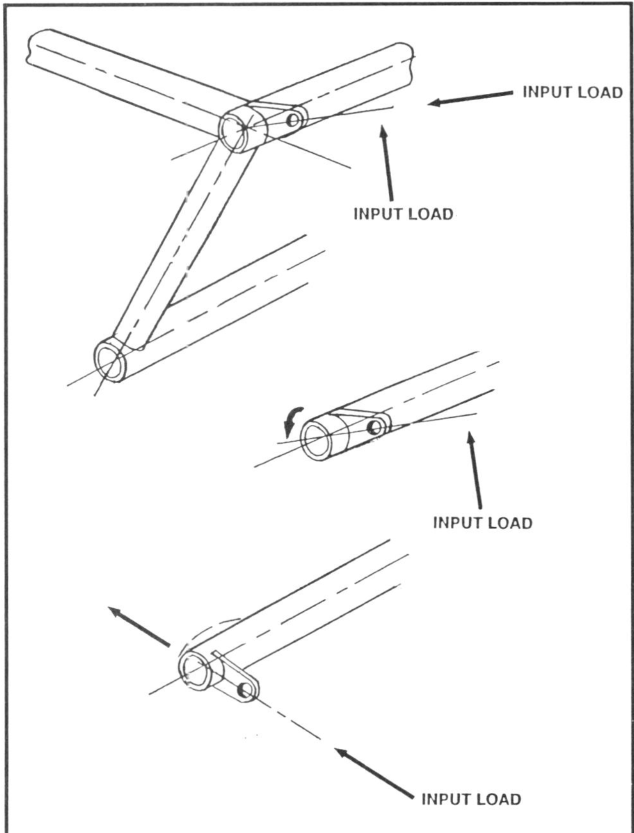

In the upper drawing, a load input in either direction places the tubes in tension and compression only because of the supporting tubes. In the lower drawing, the input load causes torsional loading of the upper tube (not good) and a bending load on the lower tube (not good either).

A light colored paint on the chassis will quickly point out any stress cracks.

This cover should be fabricated out of 1/8-inch thick steel plate.

Paint your chassis in a color other than black. Paint acts as a brittle coating which, when stressed by the underlying metal tubes, will crack when flexed or twisted. Lighter colored paints help show these stress cracks more easily. Pay attention to the paint as it will quickly display over-stressed areas which need further reinforcement or triangulation. Black paint may serve to hide those telltale cracks.

Chassis "Crash Zones"

Although the chassis and cage should be built to resist flexing and twisting, that absolute rigidity should end around the front spindle line. Forward of it should be a "deformable area" or "crash zone."

The crash zone is an area that deforms, or gives, bends or breaks, in the case of front end impact. This is done to absorb the forces of the impact and to confine the damage to as small a structural area as possible. The idea is to isolate the damaging impact forces to easily replaed components in the front before they do major damage to critical components and chassis areas.

Above and below, two different approaches to building a "crash zone" at the front of a race car. The upper car uses lighter tubes in triangulation coupled with shear bolt mounting. Below, an impact bumper assembly from a passenger car is used.

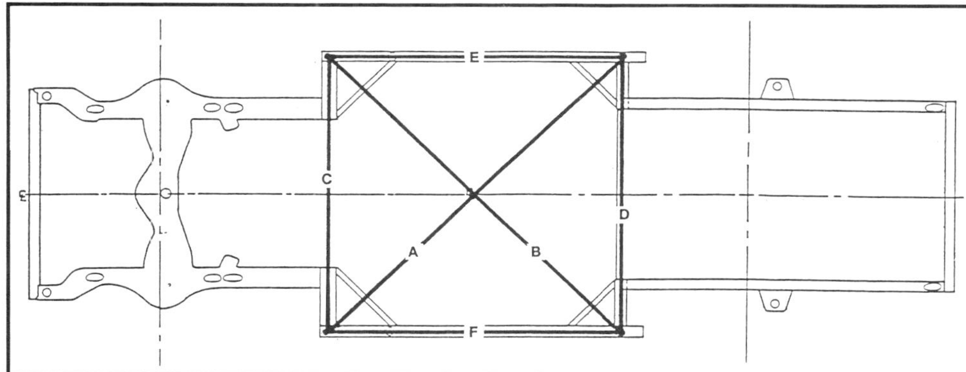

The basic reference points are four holes drilled in the chassis, one at each corner. They should be positioned so that line A equals line B, line C equals line D, and line E equals line F.

The crash zone is also vitally important for driver safety in case of a major impact. The deformable area should bend or break in order to absorb the initial shock of hitting another car hard or hitting a wall head-on. If it does not, the impact is going to be transferred to the next weakest point in the chassis — the driver. You need a cushion between the driver and the impact.

The very rear of the chassis should likewise have a crushable area to prevent major component damage and to preserve driver safety.

Chassis Reference Points

Every race car is going to endure it's share of bumping and banging against other cars and walls. And these incidents are going to take their toll on chassis straightness.

The time to start battling chassis straightness problems is just as soon as your chassis is completed — before it ever gets tweaked. When it is brand new — and straight — is the time to create reference points, and to record these. Then, when the car gets bumped or crashed, with reference points on the car and a complete written record of all measurements and angles, you can determine how to put the car back together right again, square and straight.

If your chassis was built on a jig, a perfect reference hole is the jig mounting holes in the frames.

There are a great number of ways a chassis can get bent, warped and twisted, so a number of reference points are required.

Start by drilling a reference hole in the frame at four basic points on the car. They should be at the front and rear extremes of the flat frame rails, on both sides of the frame (on ther bottom of the frame rails, for easy access). Drill them so that a measurement from side-to-side on each is the same, and a diagonal measurement each way is the same (see drawings). These four reference points will become your most important four points on the chassis. From these four points, everything else can be measured. When you do any other reference checks on the chassis, start with these four points, and always measure these first to be sure they haven't moved through any frame damage (always measure side-to-side, front-to-rear, and on both diagonals).

After your rolling chassis is built, and before the engine, driveline and body are added, it is important to establish the longitudinal mechanical center line of the chassis. This is a line that runs along the physical center of the chassis from the front to the rear. Do this by measuring center-to-center between the frame rails at the front and at the rear of the chassis. Then drill holes in the chassis at the front and the rear which will be accessible after the car is complete. Stretch a string from front to rear in line with these center holes, and take and record measurements from the string to components on each side. Measure to frame rails at several points, to A-arm mounts, to roll cage anchor points on the frame, to trailing arm brackets, etc. Be sure your measurements are accurate! After the car has been crashed, set the chassis back in the same position in the shop and re-establish this center string line, double checking all of the measurements you have recorded. This is a quick and simple way to check what is bent.

The Camaro front subframe has a hole located in the center of the crossmember under the engine. This makes a very good reference point. Record several measurements from the center of this hole to various points on the chassis to determine if the crossmember has moved, or if any other points have been moved in relationship to the

The mechanical center line of the car, front to rear, should be established, then make several reference measurements all along the chassis. Be sure you use a notebook to record all your measurements. Below, The center crossmember hole in the Camaro chassis is used for reference.

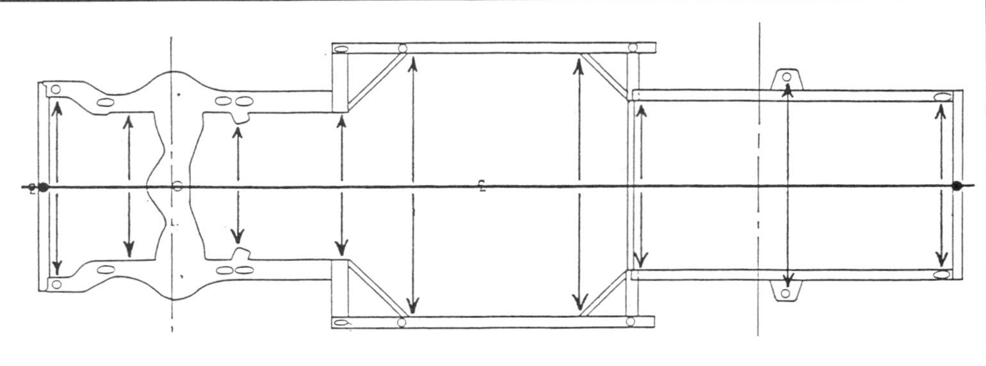

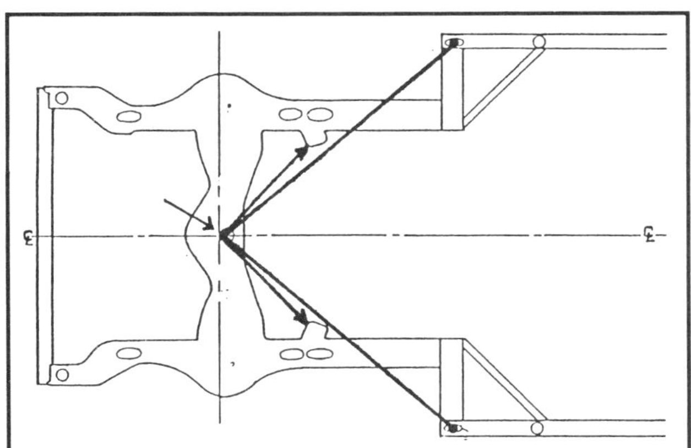

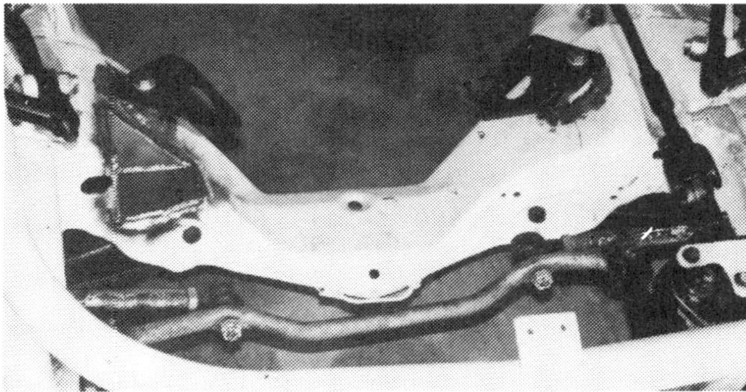

The centered reference hole in the Camaro cross member.

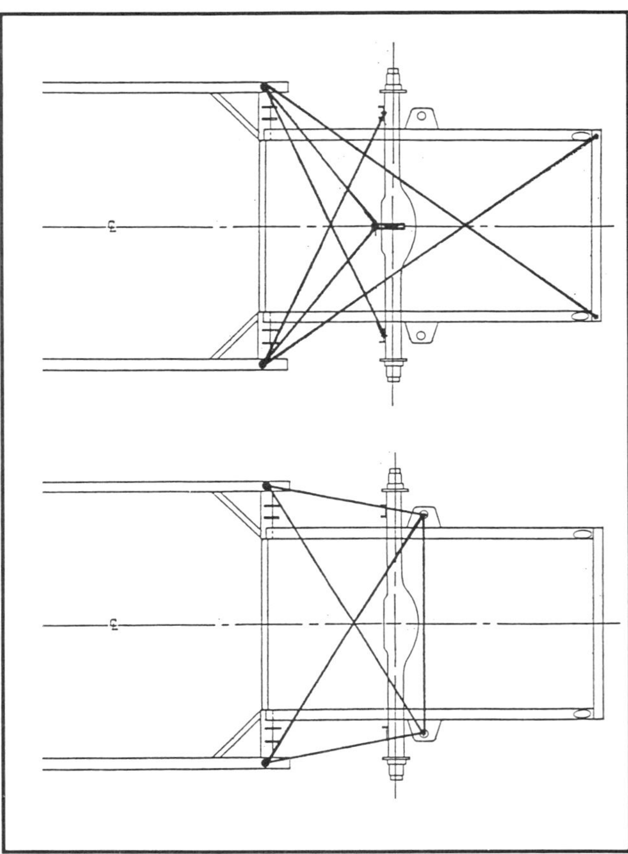

Reference measurements for the rear of the chassis

crossmember. (First be sure this crossmember hole is still on the vehicle center line, using the front to rear string.) For example, measure from the crossmember hole forward to the front frame cross piece reference hole, to reference holes at the front corners on each side of the frame, back to the corners where the subframe ties to tubular pieces that flare out to the perimeter rails, etc. Be sure you have several reference points to the upper and lower A-arm frame mounting points (both side-to-side and diagonals to other reference points on the chassis) so you can determine if these have been moved.

Going to the back of the chassis behind the rear side frame rail holes, measure to the forward chassis mounting points of the rear suspension. For instance, if your car uses a three-point rear suspension linkage, measure from the frame rail reference holes to the center of the lower arm mounting bolts on each side, and on the diagonal. In addition, measure from each frame reference hole on each side to the chassis mounting point of the upper third link. Measure from the reference holes to the center of each coil spring weight jacking bolt (and also take a diagonal

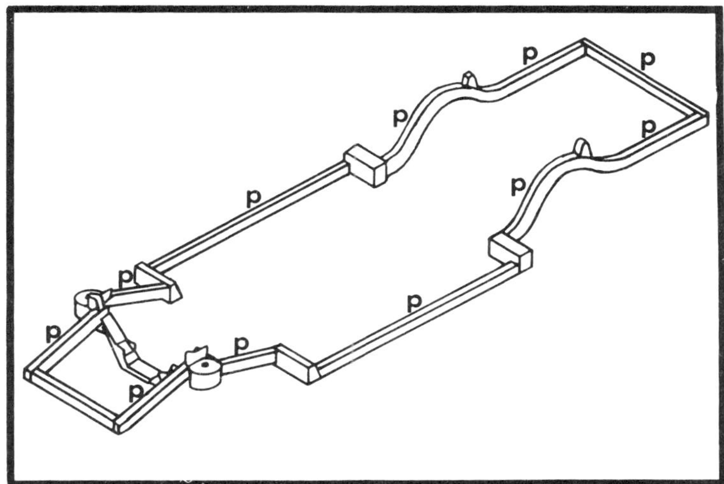

Protractor positions on the frame

measurement each way). Also measure side-to-side on the trailing arm forward mounts and on the spring weight jackers. These measurements will tell you if you have any frame compression damage.

Be sure to make reference measurements that will tell you if the Panhard bar mounting bracket on the frame has been moved forward or rearward, or has been twisted. Measure from both of your rear frame rail reference holes to the same spot on the top of the Panhard bracket, and from each of the reference holes to the bottom of the Panhard bar bracket.

The other type of frame reference measurement you need is the frame rail angles, measured when the chassis is blocked up to a specific height at a specified point on the shop floor. Use a machinist's protractor or inclinometer to measure the frame rail angle at specific marked spots along the frame on each side of the car. See the accompanying drawing for the recommended places on each frame rail to take angles. Taking these angle measurements will tell you if a frame rail has been bowed, twisted or compressed during an accident. They also give you specific reference points if you have to cut the chassis apart to weld in a new piece.

Chapter 3

Chassis Construction

This chapter contains the blueprints and construction notes for building your chassis and cage assembly. There are four versions of the car shown in blueprints, two for asphalt and two for dirt.

The basic complete asphalt track car makes use of Howe forged spindles. An alternate front stub and suspension plan is shown for the same car using the Chevrolet Impala spindles, which is a popular "wrecking yard" part for this class of car. See the "Notes On Components" subchapter later in this chapter for a complete source guide on this spindle. (This spindle differs in spindle shaft height on the upright, and in kingpin inclination. Looking at the blueprints, you will see that these two minor differences make a big difference in suspension layout.) Only the changes required in the chassis to accommodate this particular spindle are detailed.

For better overall performance, the Howe spindle is recommended. However, you may be limited by budget or track rules to using the Chevy Impala spindle. The Chevy Impala spindle is not a bad choice for the dirt track application, but on paved tracks the correct geometry is much more critical, and several changes must be made in suspension geometry and construction.

Professional chassis builders use a chassis jig. You can build a chassis in your garage without using one. Just make sure you clamp everything down tight before welding.

There are also two versions of the dirt track chassis blueprint, one for Howe spindles and one for the Chevy Impala spindles.

A Place To Build Your Chassis

The average racer building a chassis in his garage or shop isn't expected to have a surface plate or sophisticated chassis jig like professional chassis builders have. However, the racer still needs to have a flat, level and rigid reference surface which he can use to build a straight and true chassis.

Having a rigid surface to clamp the chassis pieces onto before it is welded together is extremely important. Welding two pieces together will pull them together. Weld bead will fill gaps and actually move two pieces together as they are welded. If these chassis pieces are not tied down rigidly before they are welded, the chassis will curl up like a potato chip. The chassis pieces must be clamped to a material which is much stronger than the chassis pieces being welded. For example, if you are using 3 x 3 0.090-inch wall square tubing for the main frame rails, it should

Note how everything is tightly clamped down to a rigid flat beam before welding. Pieces are tack welded first.

This is a car being built in a garage. Steel I-beams are tack welded on top of steel tube saw horses.

be clamped to at least 0.1875-inch wall 3 x 4 rectangular tubing, or an I-beam.

No concrete floor surface is absolutely flat and level. The best recommendation is to build a platform of steel rectangular tubing or steel I-beams, making a rectangular perimeter arrangement to set the chassis on, with cross pieces running across it. To level the platform, cut and weld legs from round steel tubing. You may need to do some final fine tuning to level the surface by using shims made of sheet metal and placed under the legs. The legs will also allow you to build the working platform up off the ground to a convenient working height. While this may sound like a lot of extra work and expense, it gives you a much more reliable, flat and convenient surface from which to work. (We know of one racer who built his chassis on the garage floor, supported by jack stands, with strings stretched between other jack stands for straight and level reference lines. Every time his dog ran through the garage — and his working surface — he had to reset the stands and relevel the chassis. It was a lot of extra work in the long run.)

Once the basic chassis and cage are welded together, steel tube legs can be added under the chassis for the balance of bracket fabricating and assembly.

Another approach is to tack weld round steel tubing legs to the bottom of the chassis and frame rails. When doing this, many racers make the legs the proper length to simulate the desired ground clearance of the chassis, such as four inches for the asphalt track car. The length of the legs can be fine tuned to make up for high or low spots in the garage floor so the chassis is absolutely level. (Don't do this, however, until the chassis is completely welded together, and the rear main roll cage hoop and its two main rear supports are added to the chassis. It has to be held down rigidly up through this point.)

Attaching The Front Clip and Frame Rails

To get the heights right, set the Camaro front subframe on your flat level building surface and block it so that it is level when a protractor is laid on the top frame rail near the spring bearing surface. When the subframe is in place, clamp it or tack weld it to keep it there.

Note: the blueprints show the chassis built with a reference clearance to a ground surface — 4 inches for the asphalt chassis, 5 inches for the dirt track chassis. You can locate your subframe and chassis rails above your working surface at the same height, or you can locate the chassis flat on the working surface and subtract the ground clearance from all of the measurements on the blueprints.

Next, cut the frame rails of the subframe at the point shown on the blueprint.

At this point the new rectangular tubing frame rails should be laid flat on the working surface behind the subframe, attaching the short "out" rails first and then the long main chassis rails. Tack weld the rails in place first, then put a protractor on the frame rails to be sure they are level, and measure all along the rails and at diagonals to check for

When the frame rails are laid out and ready for attachment to the Camaro subframe, make sure they are absolutely level.

squareness. Do your finish welds when you are satisfied that everything is level and square. Then continue fitting and welding the balance of the chassis frame rails.

Building The Rest of the Chassis

Modification of the front spring bearing surface in the frame is next. Cut the stock spring pocket and A-arm tower out of the frame. Replace that with a flat 0.185-inch thick steel plate cut to size for the hole. Make sure that this plate is absolutely level when welded in as this will become an important chassis reference surface.

In the center of this plate a 1-3/16-inch diameter hole is drilled for the weight jacker screw. Be sure that the hole for the weight jacker is centered over the diameter of the spring. Keep in mind that the mounting angle for the spring on the stock Chevy and Camaro lower A-arm angles the spring inward about 5 degrees. So mount the lower A-arm and spring to find the exact location of the center of the spring at normal ride height position. After the hole is drilled, the weight jacker screw bolt can be tacked into place. After it is in place, re-attach the lower A-arm and spring and install the weight jack and spring retainer just to be certain everything is correct. The complete weight jack, bolt (with grease fitting) and spring retainer assembly can be purchased from PRE under part number 02-095. The screw is fitted with a 1/2-inch square drive at its head.

Next attach the top A-arm mounting bracket. This piece can be cut from 3/8-inch thick steel plate. The upper A-arm bracket must be mounted in an axis plane which is parallel to the lower A-arm mounting plane, which is 20

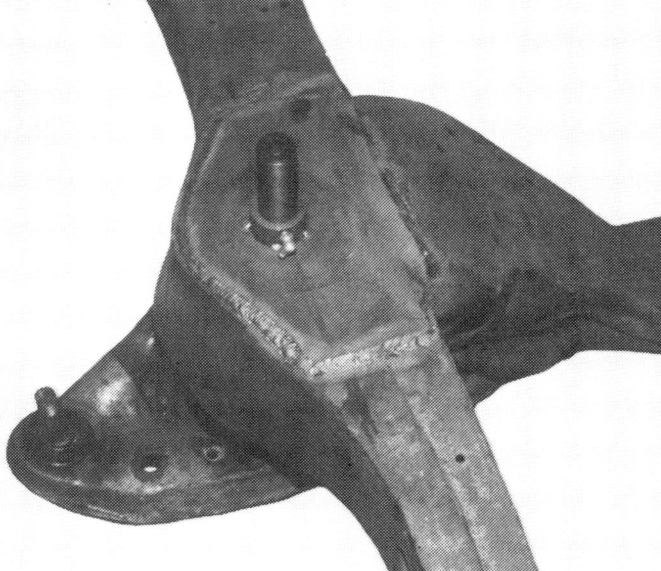

The first step with the subframe is to cut out the top of the spring pocket and replace it with steel plate. Then locate the center of the spring and locate the weight jacker assembly.

PRE offers this piece which is a combination spring pocket top plate and upper A-arm mount. The weight jacker hole is already installed too. The A-arm bracket is drilled for a 7.75-inch center spacing cross shaft, but it can be redrilled to accept the Ford-based 6-inch shaft. The PRE bracket is shown below in place on a PRE-built chassis.

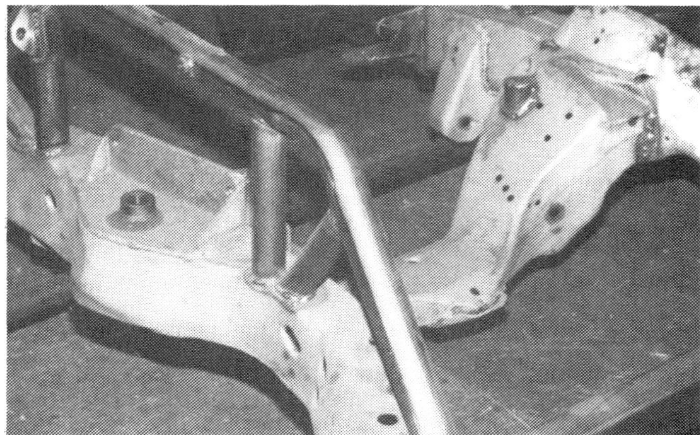

The upper A-arm must be attached at an angle parallel to the lower A-arm swing axis.

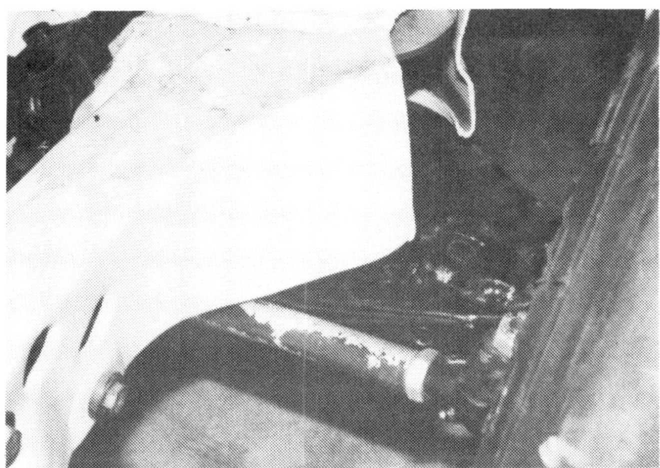

Above, note the frame has not been notched for tie rod clearance. This can provide for some disastrous consequences. Also note how the spring pocket has been notched for shock absorber clearance so the shock can be mounted just behind the lower ball joint. Below, note how this frame has been properly notched for tie rod clearance. Also note how every weld on the frame has been rewelded, and the upper shock mount is a straddle mount bracket for extra strength.

degrees from straight ahead. Even though the blueprint shows a position for mounting this bracket, there is only one sure way to get it in the proper position. That is, mount all the suspension pieces, then position and measure.

Attach the lower A-arm with ball joint installed to the chassis, the spindle and hub to the lower A-arm, the upper A-arm to the spindle, and the upper A-arm to the pre-cut mounting plate. The lower A-arm must be positioned for its proper running height so the lower ball joint pivot point is the correct distance from the reference surface. The spindle is then positioned for an initial pre-set caster and camber using a dial protractor set against the hub and spindle. For the right front spindle, the camber should be set initially at negative 1-1/2 degrees, and the caster at positive 2 degrees. The left front spindle should be set for 1 degree positive camber and 0 degrees caster. Secure the spindles in their correct position so they cannot move (you might want to tack weld a small bracket to the spindle upright). Then tack weld the mounting bracket in place. Double check the caster and camber settings again before you do the final welding on the mounting bracket.

Check the blueprints and photos that indicate areas on the Camaro front subframe which must be trimmed for proper component clearance. Also note that the outside vertical center section of the spring pocket must be trimmed away in order to accommodate the shock absorbers. The best mounting position for the front shock absorbers is to run them up inside the upper A-arms. This allows the shock absorber to be mounted just inside the lower ball joint for better shock control. Many commercial chassis builders do not mount their shocks like this because it is a more time consuming modification of the frame. Cutting the frame section in the spring pocket like this does not weaken the chassis structure in any way.

Finish the rest of the attaching brackets on the chassis, then begin the roll cage. The separate roll cage blueprint has numbers on it which indicate the order in which the cage members should be attached.

A very convenient way of working on a chassis is by attaching it to a rotating chassis stand. Plans for building one are included in "The Racer's Guide To Fabricating Shop Equipment" from Steve Smith Autosports.

Notes On Components

Front Subframe

The front subframe to use is the 1971 to 1981 Camaro or Firebird. The reason for the overwhelming popularity of this frame section has already been detailed at the beginning of the previous chapter. The 1975 to 1979 Chevrolet Nova subframe is similar to the Camaro. It is identical in every way, up to the frame horns in front of the front spring pockets, where the frame horns come in different shapes depending on year and model.

Lower A-Arms

The 1971-1981 Camaro and 1975-1979 Nova lower A-arms are dimensionally the same as the Chevrolet Impala and other GM big car models, except for the size of the lower inner bushing holes and the lower ball joint hole. The Impala and GM big cars use a larger ball joint (Moog K-6141), which is more desirable for a racing application. This ball joint fits both spindles we have shown with our blueprints, the Howe and the Chevy Impala. So the best plan is to use lower A-arms from the GM big car group, which is as follows:

1973-1976 Chevrolet Impala, Caprice, Belair
1973-1976 Buick LeSabre
1973-1976 Oldsmobile 88 series
1973-1976 Pontiac Catalina, Bonneville, Grandville

Spindles

We have shown in our blueprints two different spindles, the Howe forged spindle, part number 344GN, and what we have referred to in this book as the Chevy Impala spindle. This spindle is found in the same GM big car group models as shown above for the lower A-arms. The Chevy Impala, Howe forged and PRE Ultra Drop spindles all accept the same upper ball joint (Moog K-6024), lower ball joint (Moog K-6141) and hub, rotor and bearings.

When buying a wrecking yard spindle, be sure to check it out thoroughly to be sure it is not damaged. Check for: spindle shaft straightness, true kingpin inclination angle, cracks at the bearing shoulders, cracks in the steering arm at the tie rod hole, cracks at the bend leading to the upper ball joint hole, and damaged or distorted ball joint holes (which is a sign of severe impact or severe misalignment and wear).

Lower A-Arm Bushings

Use the PRE three-piece steel and Nylatron bushings. They are available in six different sizes to work in whatever GM lower A-arms you choose to use. This bushing has a steel outer with a steel inner piece that rides on a Nylatron liner. Nylatron is a space age material somewhat like nylon except that it will not deform. It is a high strength material with a dampening quality, has high fatigue and impact resistance, natural lubricity and resistance to high temperatures. This special bushing liner prevents steel to steel contact with the inherent problem of galling and siezing (which is a big source of suspension handling problems).

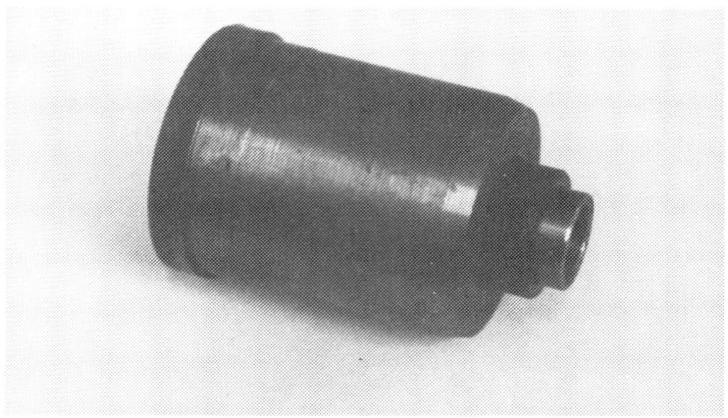

This 3-piece lower A-arm bushing from PRE features a Nylatron insert that prevents steel to steel contact and thus siezing and galling.

Ball Joints

With any of the spindles we have specified, use the Moog K-6024 for the upper ball joint and the Moog K-6141 for the lower ball joint.

Tie Rods

Steel tubular A-arms can be constructed from 4130 steel tubing, 1-inch O.D., 0.156-inch I.D., 14 inches long. Thread one end for left thread, one end for right thread, using 5/8-18 thread. Use a quality 5/8-inch spherical rod bearing at each end instead of a tie rod end. Mount to the steering arm with a spacer between the rod end ball and the steering arm to insure proper rod end travel. The length of the spacer will be one of the adjustments for bump steer.

When you use spherical rod end bearings in your steering system, make sure you use only the best aircraft quality rod ends, not standard grade rod ends. They are more expensive, but safety in your steering system is very important too. AFCO has available rod ends that are especially made for use in steering systems. This rod end has a 5/8-inch bore and a 5/8-inch thread shank, but the body is the size and thickness of a 3/4-inch rod end. It is available with left hand thread shank (part number 10401) and right hand thread (part number 10402).

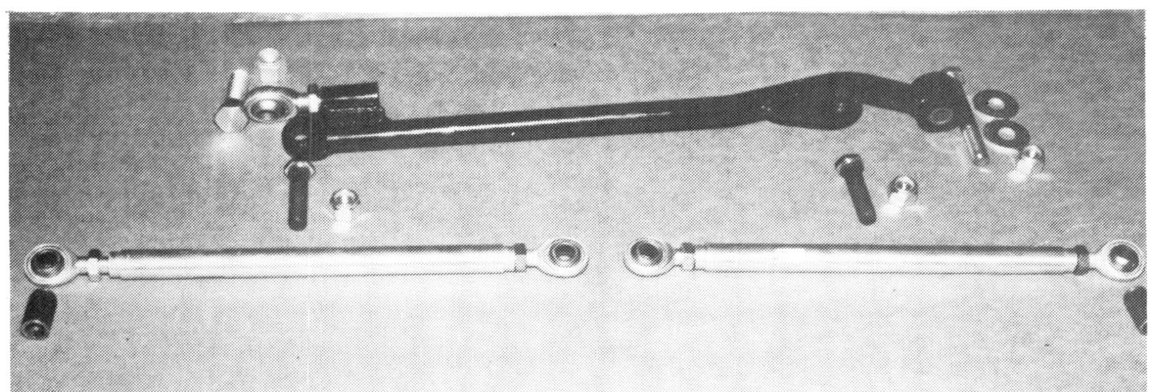

Left the PRE steering drag link and tie rod kit.

Below, the drag link installed with a power steering box.

Steering Drag Link

PRE has available a unique adjustable bump steer drag link. Gary Sigman of PRE worked for six months experimenting with different drag links, cutting and welding, heating and bending, trying to find a way to effectively modify a stock component. No stock part worked out right for the bump steer and clearance. So he designed and built his own. It is made from chrome moly steel tubing and uses a rod end bearing and polyurethane bushing. It is designed so you can get the desired bump steer through five inches of bump travel. It is made to be used with a Camaro-based power steering unit. See the Front Suspension and Steering chapter for more information on this power steering.

If you want to go the stock component route, you can use the Moog DS-830 drag link, and modify it to accept the spherical rod end bearings from the tie rods (tie rod holes must be drilled to accept a 5/8-inch bolt). Be aware that the tie rod inner mounting points will not be in the correct position for working out the bump steer.

Upper A-Arm Cross Shaft

The most popular upper A-arm cross shaft which has been used for years to build upper A-arms is a Ford-based part. The mounting holes are spaced 6 inches center to center with 9/16-inch diameter holes. The application source of these cross shafts are as follows:

1965-1972 Ford and Mercury — all large body pasenger cars

1967-1972 Thunderbird

1969-1970 Lincoln

1972 Ford Torino

This cross shaft is also found in the Moog upper control arm kit number K-8175 (the kit also contains parts that you won't use). The Moog number for the shaft itself is 8385.

Some racers use a cross shaft that is a 7-3/4-inch center to center bolt pattern. The application is 1971-1973 Camaro and Impala, and some other GM cars. But this wider cross shaft takes up too much room in a cramped area. Another design element of this cross shaft that makes it popular is that it features a 1/2-inch offset between the bolt mounting surface and the center line of the A-arm bushing pivot. This means the cross shaft can be flipped over to lengthen or shorten the A-arm by 1/2-inch and make it more universal in its application.

Construction Blueprints

The following pages contain the chassis building blueprints plus photos of various detail items. The photos are included to help clarify some areas which are difficult to discern from the prints.

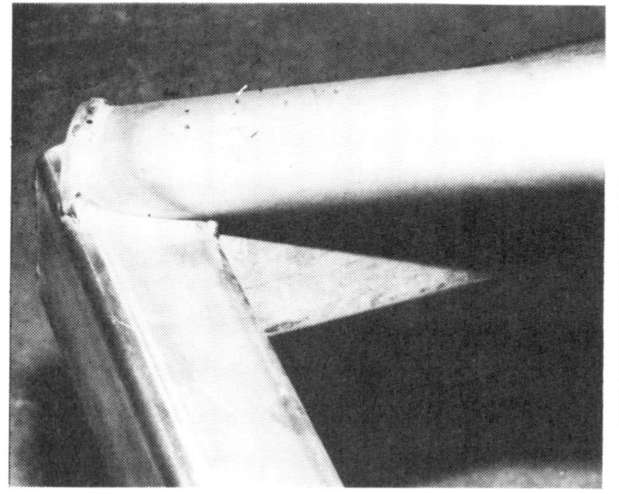

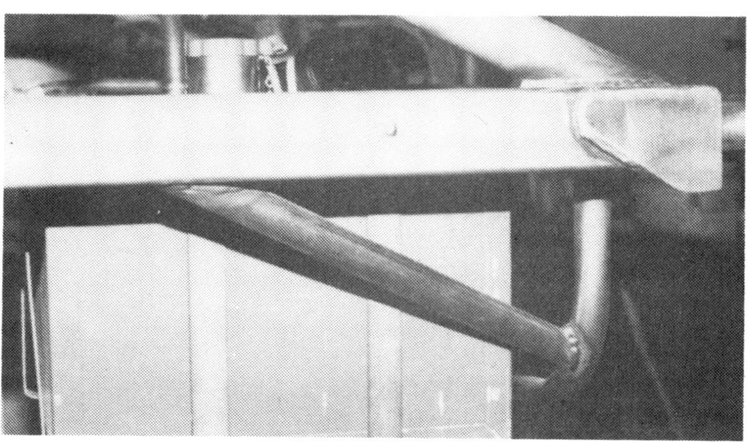

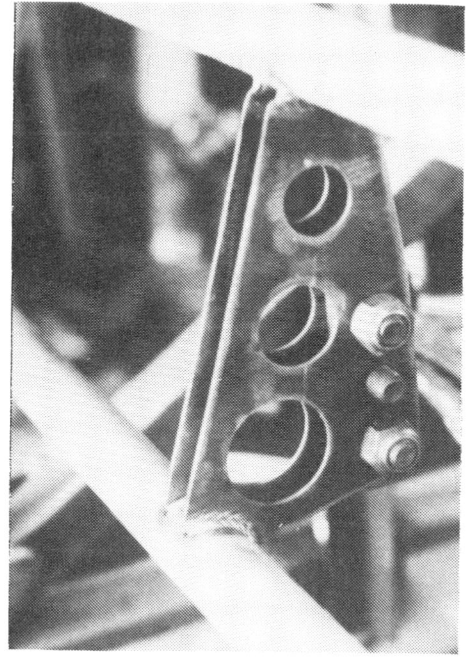

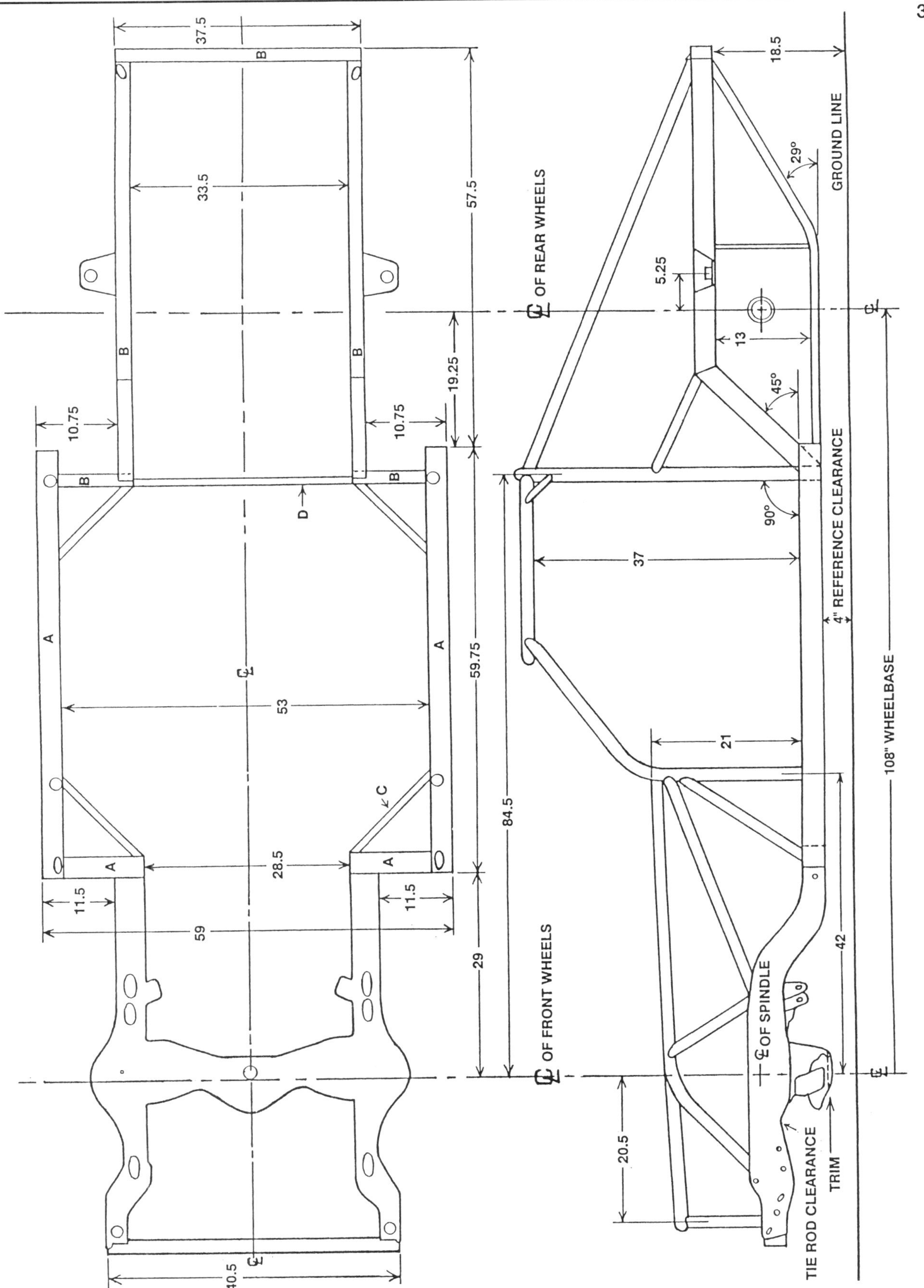

Asphalt chassis basic frame and cage layout with Howe spindle being used. Letters relate to materials. See materials legend, page 40.

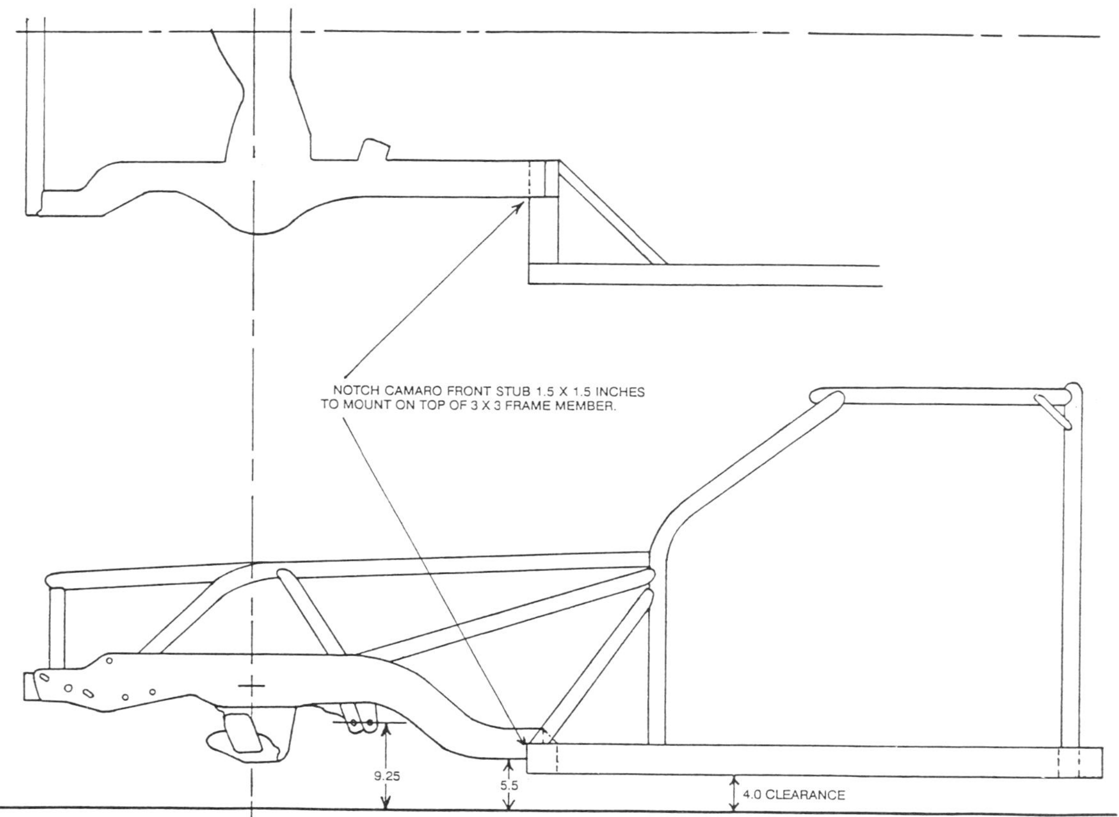

Asphalt car chassis modification for use with Chevy spindle.

MATERIALS LEGEND

A	3 X 3 X .095 WALL SQUARE TUBING
B	2 X 3 X .095 WALL RECTANGULAR TUBING
C	1.5 OD X .095 WALL ROUND TUBING
D	1.25 OD X .095 WALL ROUND TUBING
E	1.75 OD X .095 WALL ROUND TUBING
F	1.5 OD X .083 WALL ROUND TUBING
G	.1875 THICK STEEL PLATE (BRACKETS)
H	1.75 OD X .083 WALL ROUND TUBING

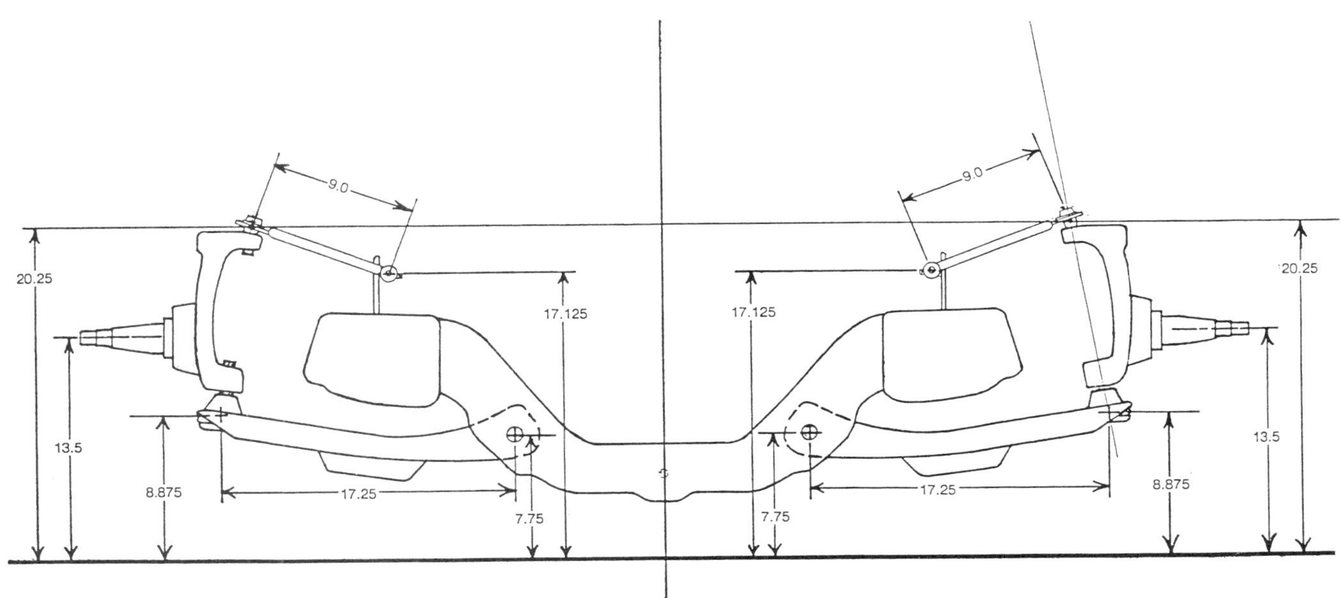

Asphalt chassis front end layout using Howe spindles. 4-inch front roll center, 64.5 track width.

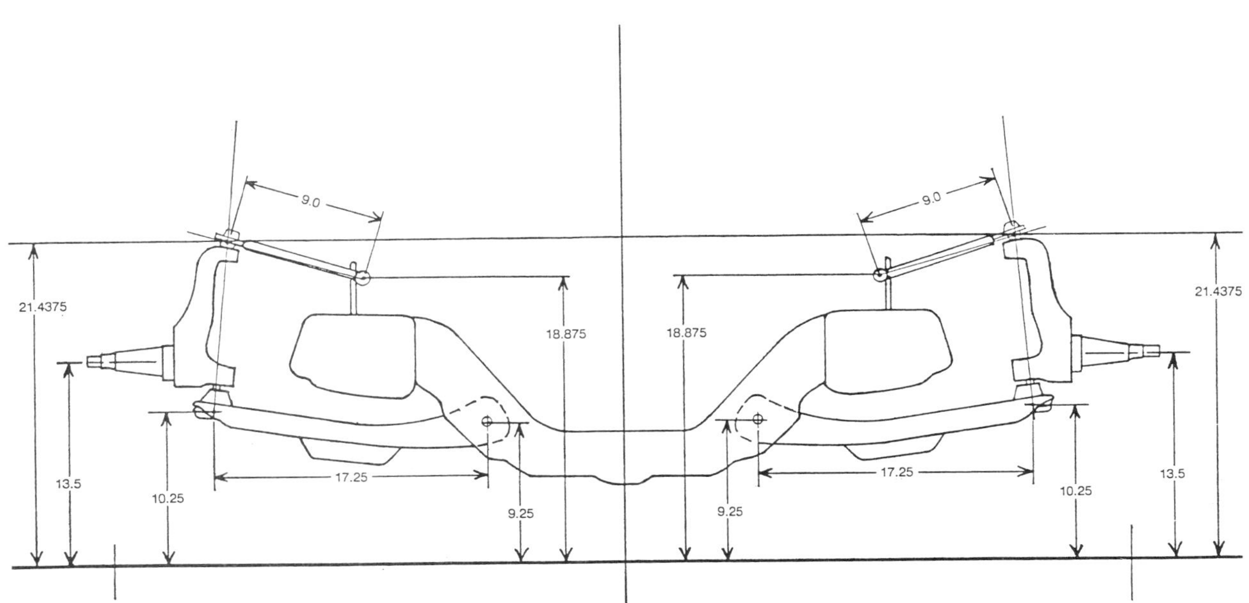

Asphalt chassis front end layout using Chevy spindles. 4-inch front roll center, 64.5 track width.

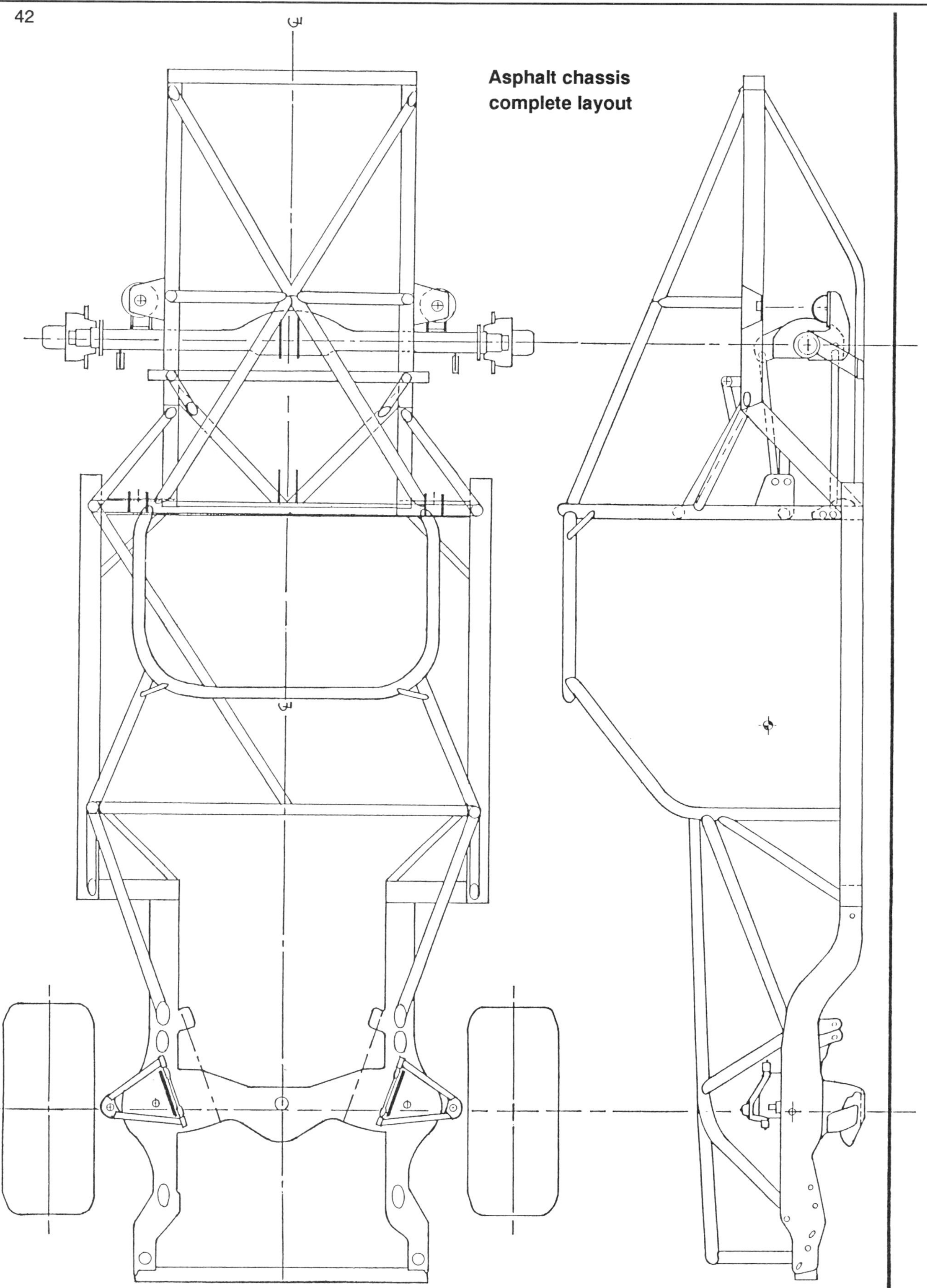

Asphalt chassis complete layout

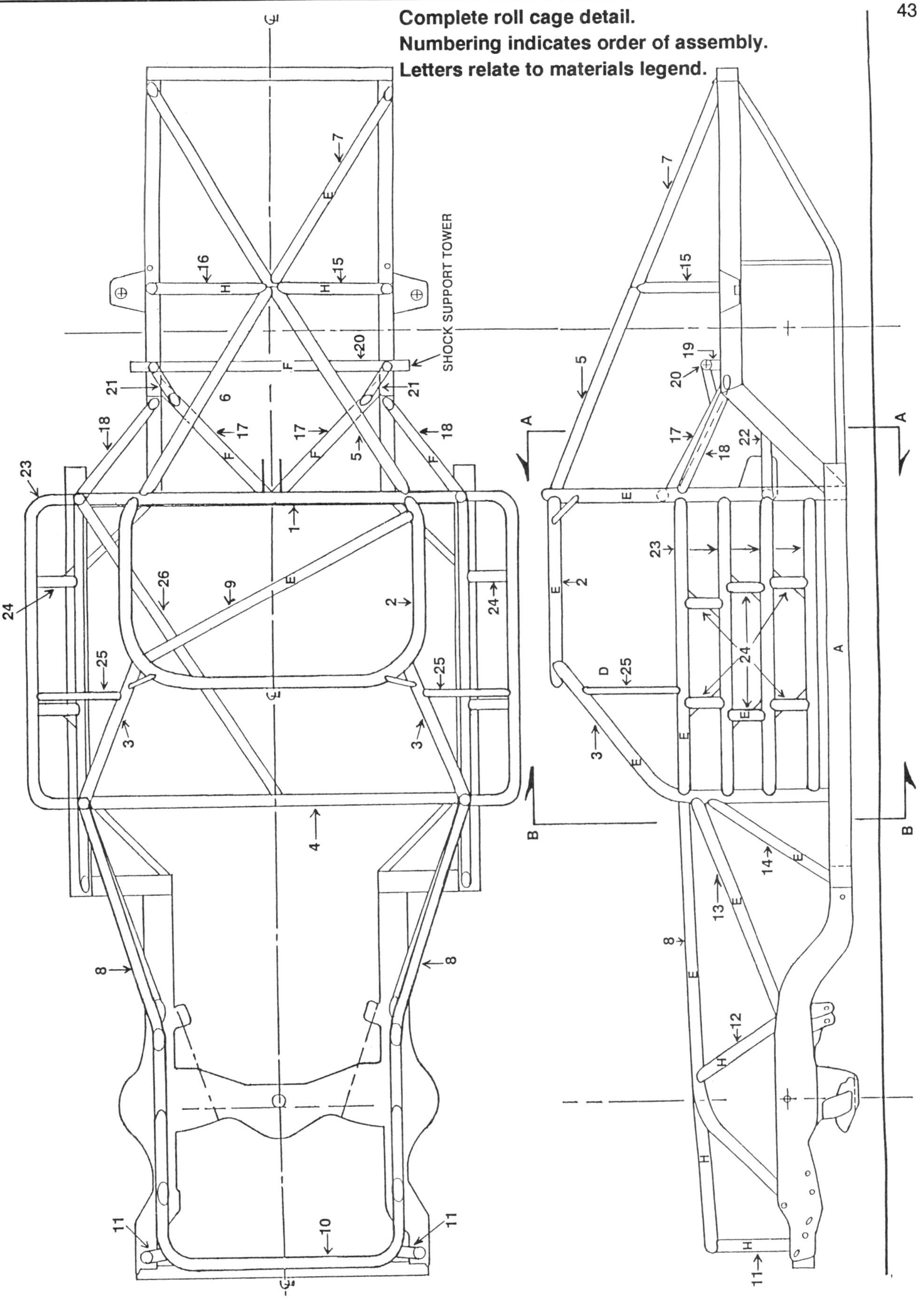

Complete roll cage detail.
Numbering indicates order of assembly.
Letters relate to materials legend.

SHOCK SUPPORT TOWER

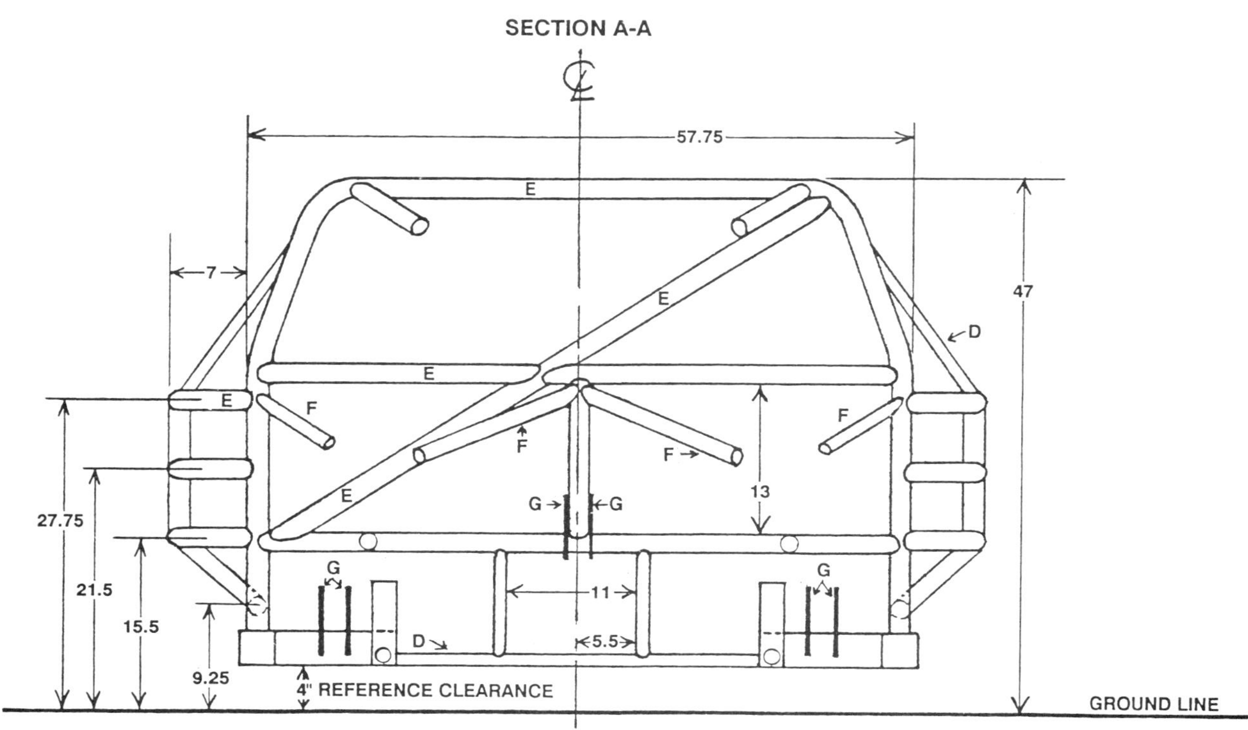

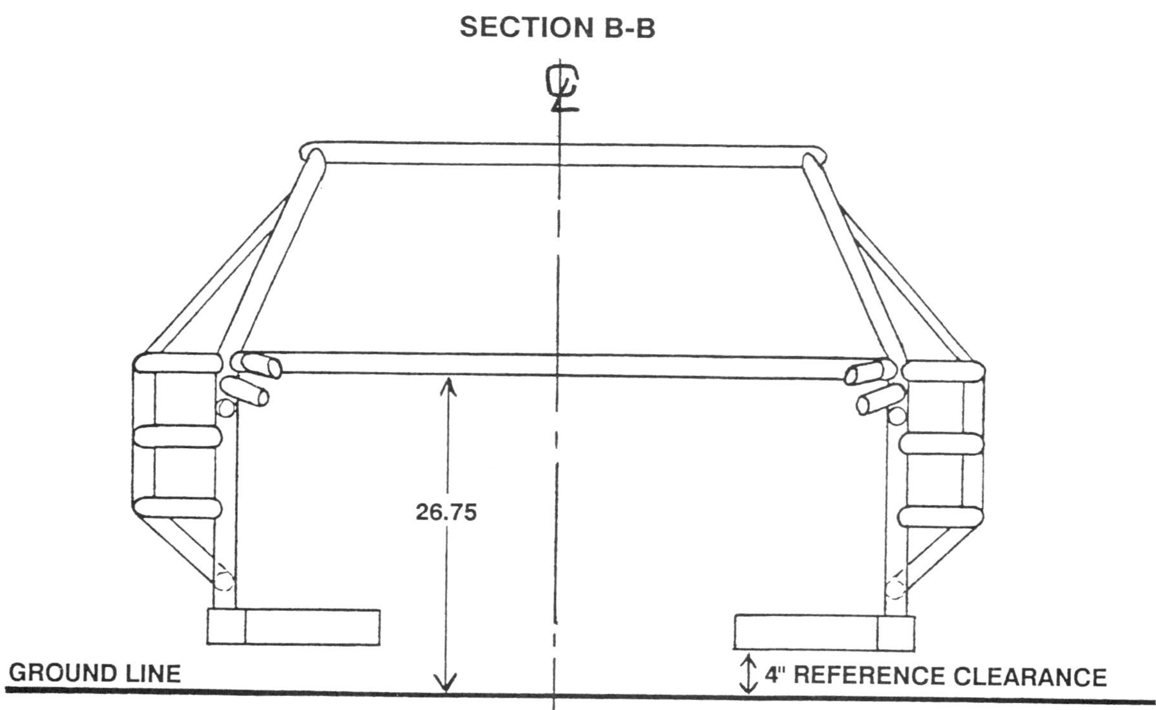

Roll cage cross section details

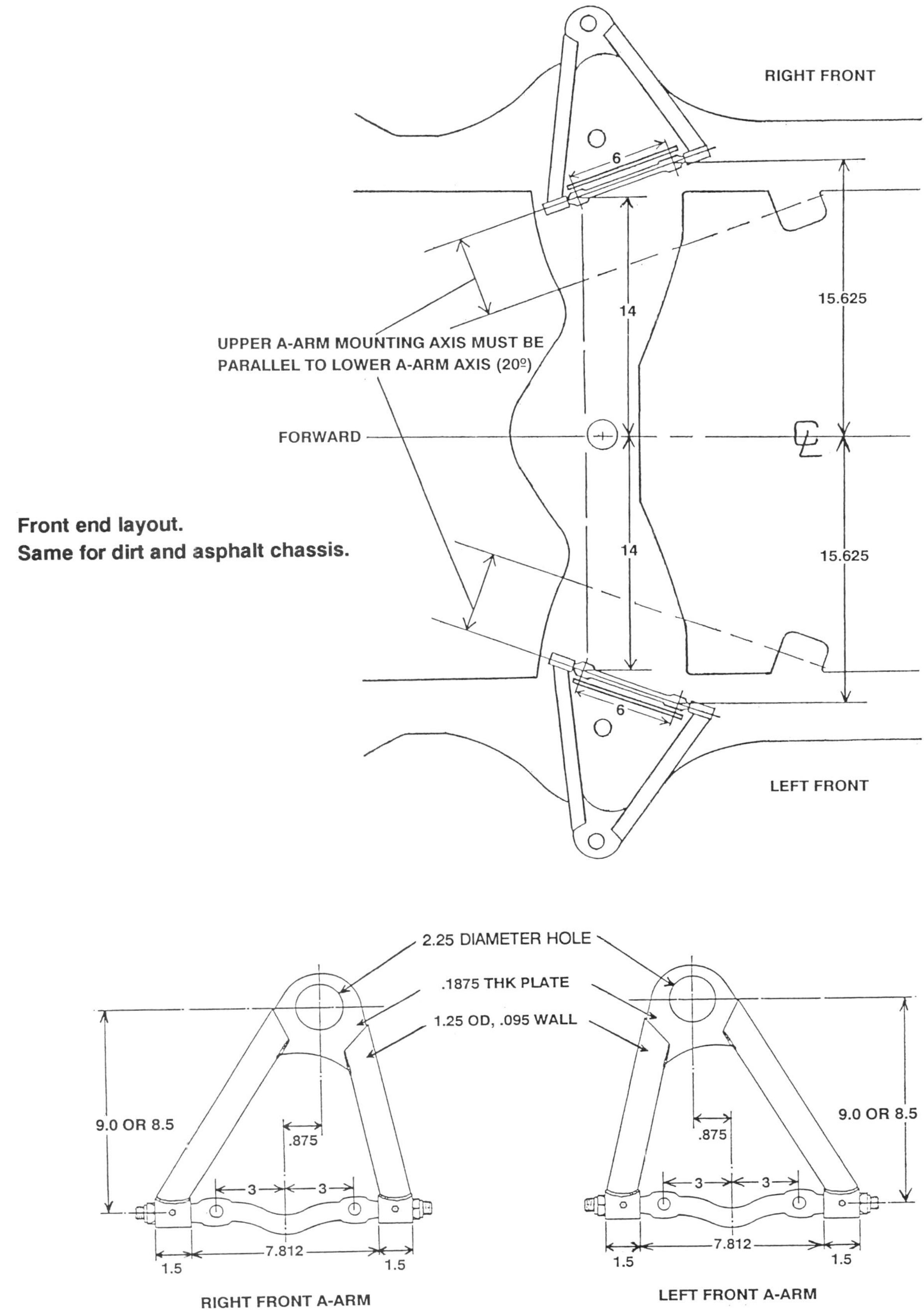

Front end layout.
Same for dirt and asphalt chassis.

A-arm detail. 9.0-inch length is for asphalt chassis, 8.5-inch length for dirt chassis.

ASPHALT CHASSIS REAR SUSPENSION DETAIL

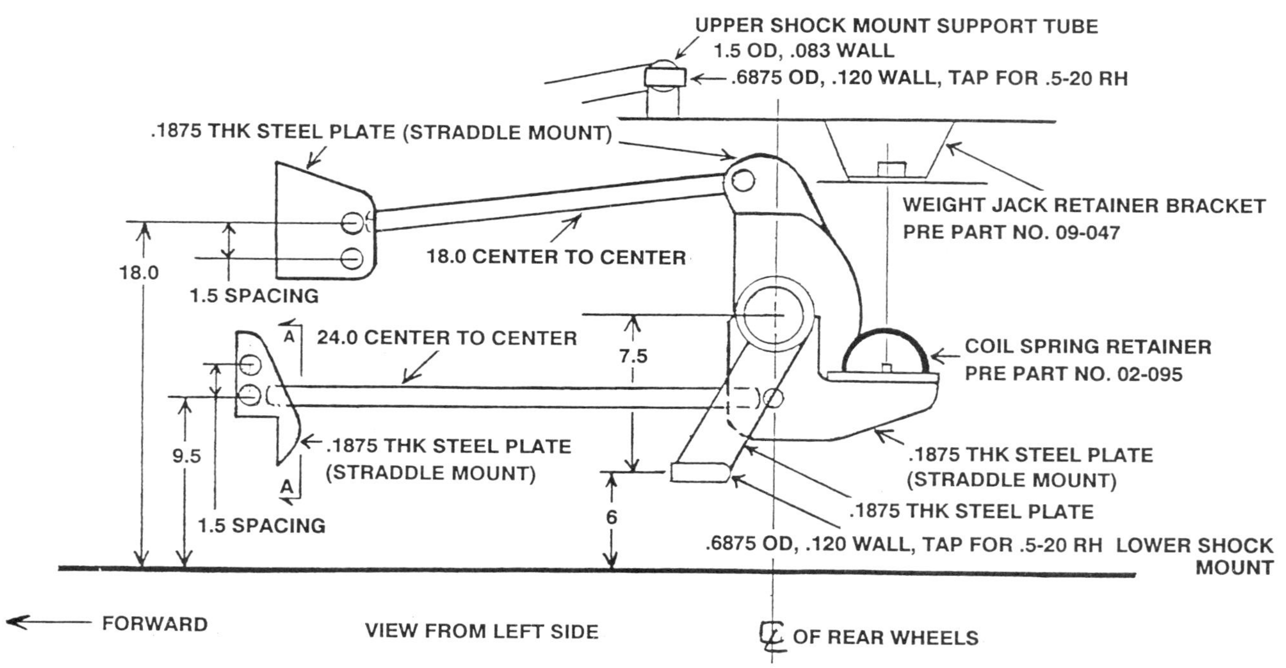

SECTION A-A

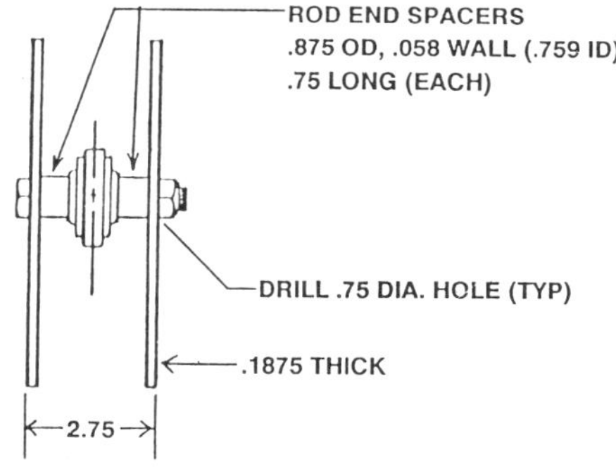

LOWER LINK DETAIL

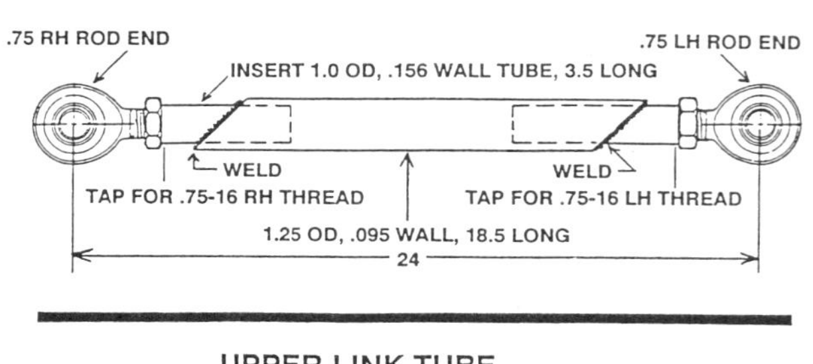

UPPER LINK TUBE

1.125 OD, .219 WALL. 15.0 LONG
TAP ONE END FOR .75-16 RH THREAD
TAP ONE END FOR .75-16 LH THREAD

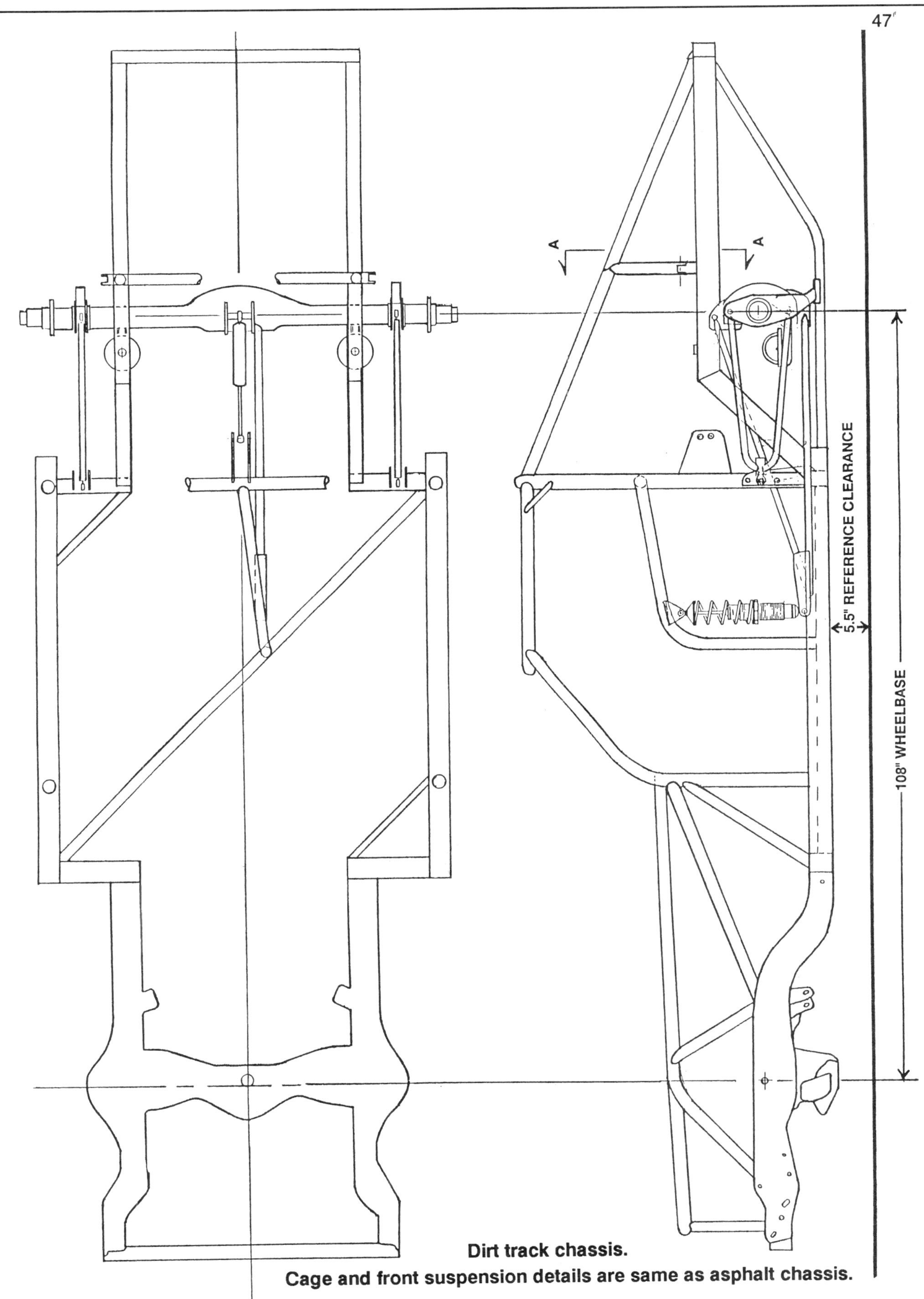

Dirt track chassis.

Cage and front suspension details are same as asphalt chassis.

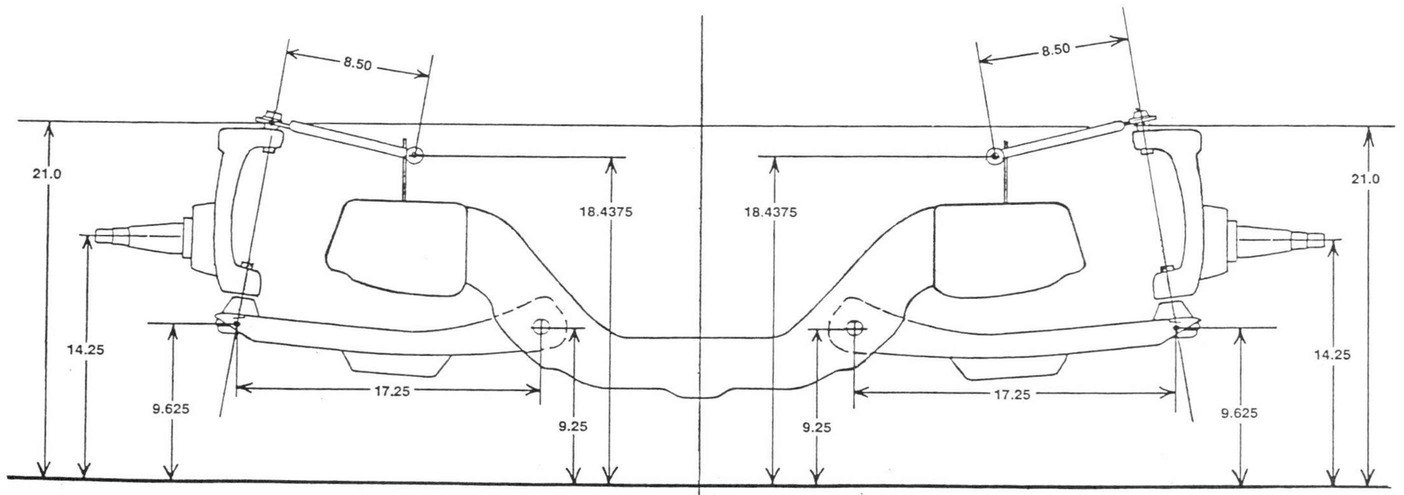

Dirt chassis front end layout using Howe spindles. 4-inch roll center, 64.5 track width.

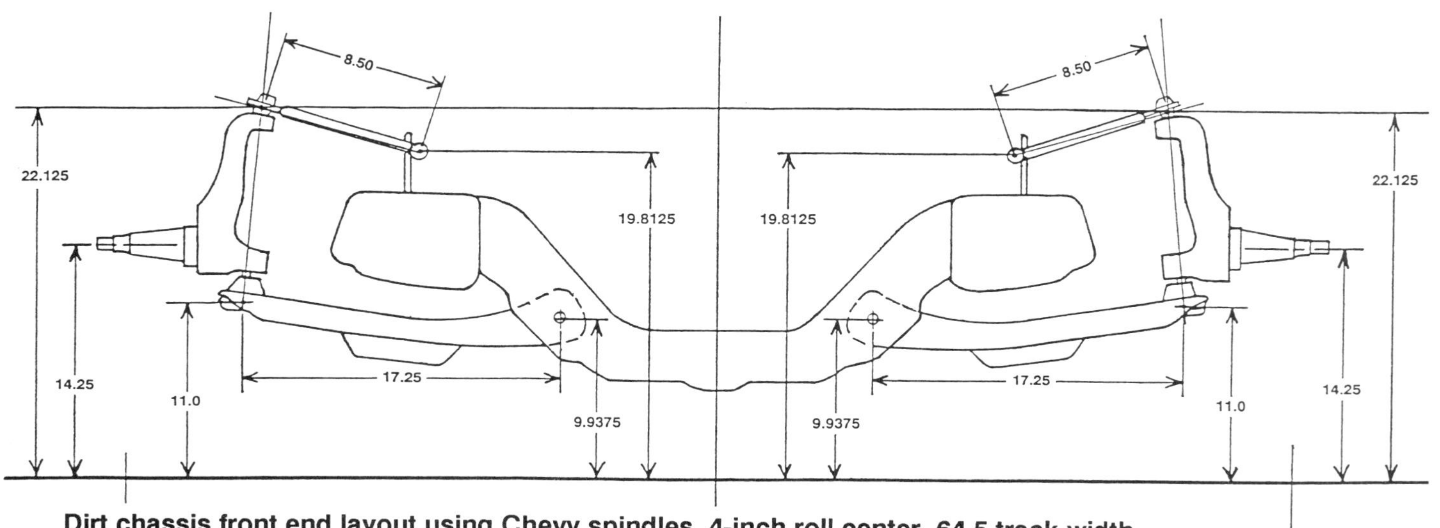

Dirt chassis front end layout using Chevy spindles. 4-inch roll center, 64.5 track width.

SECTION A - A

℄

SHOCK ABSORBER MOUNT AT END OF TUBE

NOTCH END OF TUBE FOR STRADDLE MOUNT BRACKET

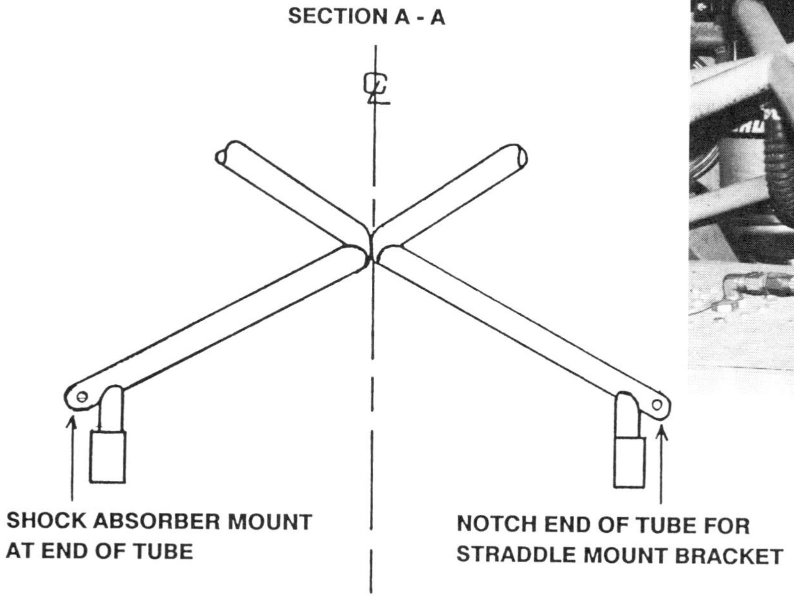

FLOATER BRACKET DETAIL

SIDE VIEW

BACK VIEW (RIGHT SIDE)

FLOATER TUBE MATERIAL 3.25 OD,.120 WALL (3.010 ID)

MAKE RETAINER RINGS
FROM SAME MATERIAL

ROD END SPACERS
.875 OD, .058 WALL (.759 ID)
.25 LONG

1.75

BRAKE BRACKET

DRILL .75 DIA. HOLE (TYP)

.1875 THICK

8

8

SHOCK

4.5

5

.010 TO .015 CLEARANCE

INSTALL GREASE FITTINGS
FRONT AND REAR

SHOCK MOUNT
.6875 OD, .120 WALL
TAP FOR .5 - 20 RH THREAD

WISHBONE LINK DETAIL

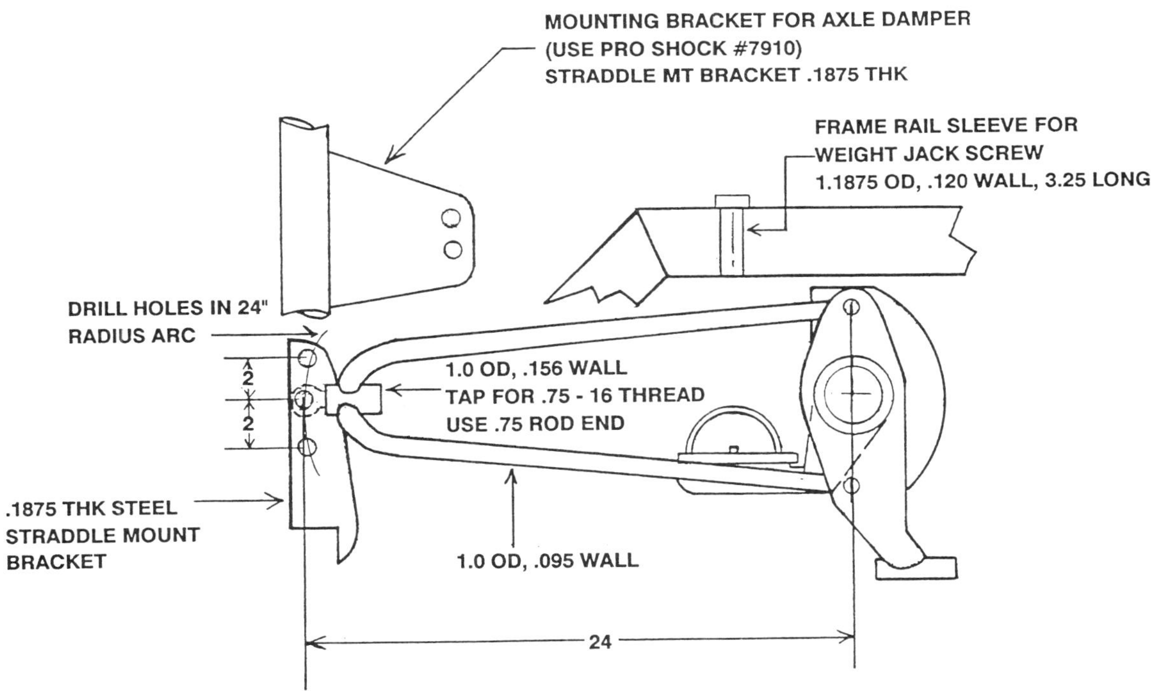

TORQUE ARM SYSTEM

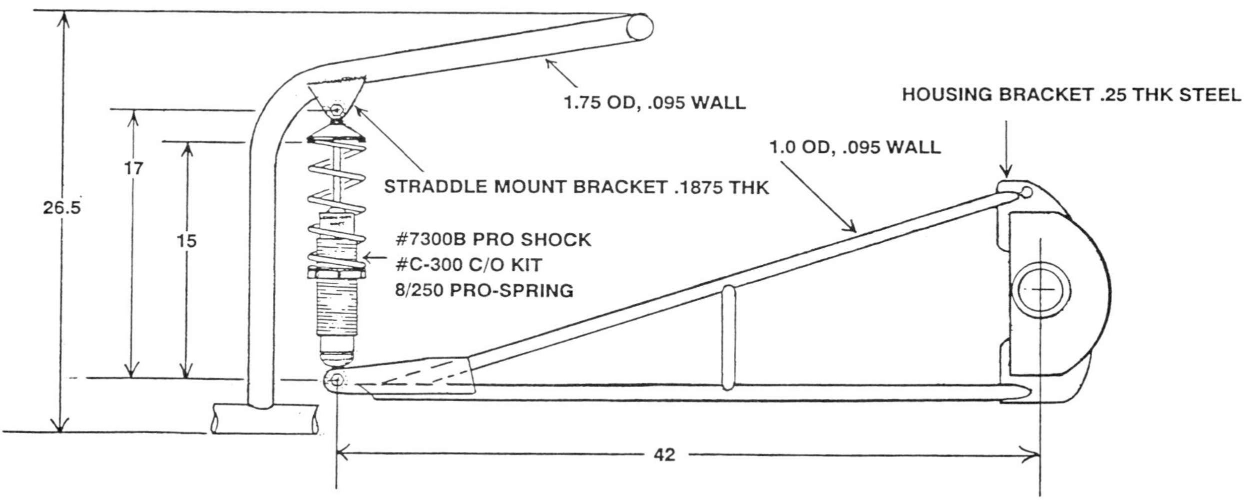

Chapter 4

Front Suspension and Steering

Camber

Camber is defined as the inward or outward tilt of a wheel at the top relative to vertical at the center of the wheel in the lateral plane. Zero camber is then true vertical, negative camber is the tilt of the top of the wheel toward the center of the vehicle, and positive camber is the tilt of the top of the wheel away from the center of the vehicle. If the top of the tire is leaning inward toward the center of the car (being viewed from the front of the vehicle), the tire has negative camber. Conversely, if the top of the tire is leaning outward away from the center of the car, it has positive camber.

The proper range of camber adjustment is discussed in the Chassis Adjustment chapter. Using tire temperatures to fine tune maximum performance with camber settings is discussed in the Tires chapter.

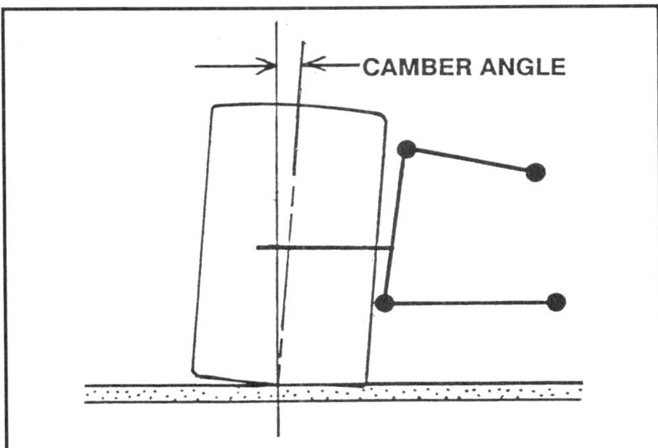

Tracking Camber Change

If you want to find out the true dynamic camber change curve of your car, DON'T block the car at ride height and move the right front wheel into bump, measuring the camber at 1-inch increments. This is not correct — it just simulates one wheel bump, and not body roll. It doesn't take into consideration the positive camber gained at the right front due to body roll. In one-wheel bump, the A-arms pivot about their instant center, and not the roll center.

To simulate camber change during body roll, jack the car up on the left side at the lateral CG so that the inner pivot points of the A-arms move downward. Now the car is rolling about the roll center. Have some type of friction limiting plate under the right front tire — such as a piece of cardboard — because the tire is going to want to move laterally somewhat. If you prevent this lateral movement, you will get false readings.

Caster

Caster is the inclination of the steering axis from vertical in the longitudinal plane (wheel viewed from the side). Positive caster is achieved when the steering axis is inclined toward the rear of the vehicle at the top in the side view. Negative caster is achieved when the steering axis is inclined toward the front of the vehicle at the top in the side view.

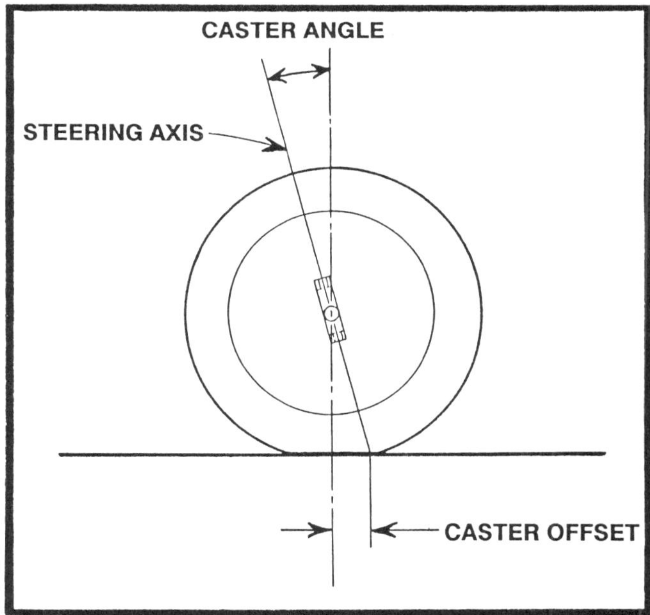

Positive caster provides a self-centering effect to the steering (called self alignment torque). The more positive caster, the more tracking feel there is for the driver. However, the more caster used, the harder the driver's steering effort becomes.

Negative caster creates the opposite of a self aligning torque. Where the positive caster tries to resist a steering effort and align the wheel straight, negative caster will help the wheel to deviate from a straight path. Negative caster, then, could almost be said to act as a form of self-energizing power steering.

Caster stagger is the difference between the amount of left front and right front caster. When caster angles are not equal, the steering will tend to pull toward the side with the least amount of positive caster or the side with more negative caster. On stock cars on an oval track, a greater amount of right front positive caster will enable the driver to steer the car into a corner with less effort. The greater the amount of caster angle between the left front and the right front, the greater the caster stagger.

Caster stagger can be anywhere from 1 to 4 degrees. Short tight tracks can use more caster stagger to help the car make a tight arc with minimal effort. However, more caster stagger makes it much more difficult to turn the car back to the right. This makes it hard to maneuver in traffic and to drive out of trouble. Power steering helps minimize the problems associated with more positive caster and more caster stagger. The bottom line on stagger: make it comfortable for the driver.

Toe-Out

Toe-out is the difference in distance between the front and rear measurements of the tires on the same axle in the center of the tread surface, at spindle height. where the front measurement is greater than the rear. Toe-in is the opposite.

Toe-out is a static alignment made to minimize tire scrub and rolling resistance which develop through the tires when a car is cornering. When a car is in a turn, both front wheels are turning about a common center, but because the left front is closer to that center, it is turning on a sharper radius. Thus with the two front wheels of the car turning on separate radii, the two tires are actually pointing in different directions with the left wheel making the sharper turn. The purpose here is to introduce more initial toe-out so each front wheel follows its own radius in a perfect path through the turn. When either front wheel is at an angle against its turning radius, it is requiring extra horsepower to scrub that tire through the turn.

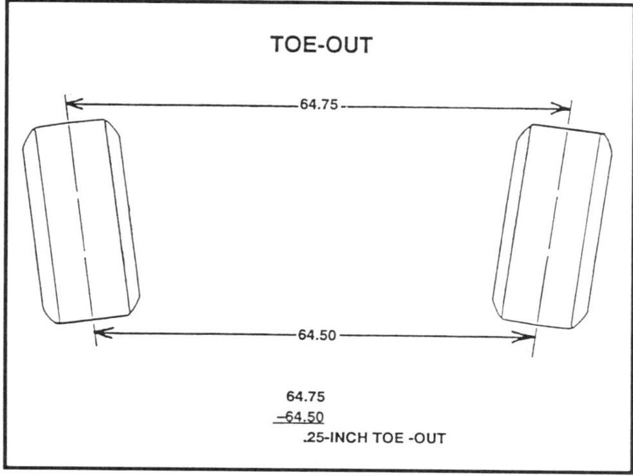

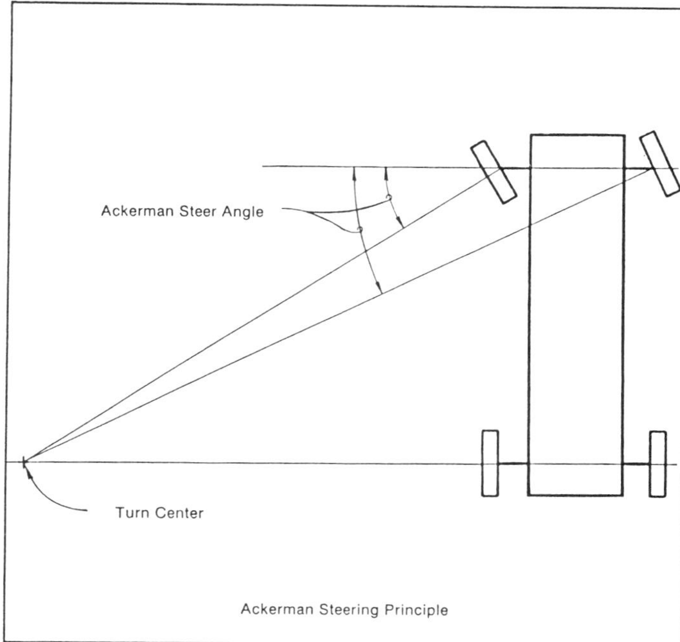

Ackerman Steering Principle

Ackerman Steering

Ackerman steering geometry is where the inside front wheel is steered in a sharper arc than the outside front in order to eliminate tire scrub at the inside wheel. For a number of years, it was thought that Ackerman steering wasn't important to race cars because of tire slip angles under hard cornering. Parallel steering and even anti-Ackerman steering was in vogue for a while. But now current thinking and advanced technology have changed all that.

TODAY, the current technology has the left front tire much more heavily loaded, which means it participates much more in the steering and handling of the race car. The left front is now a very important tire contact patch. And thus, the direction it steers without tire scrub is now very important. It must steer at a greater arc than the right front because it is travelling on a shorter radius. The left front simply steers more and helps point and stabilize the car.

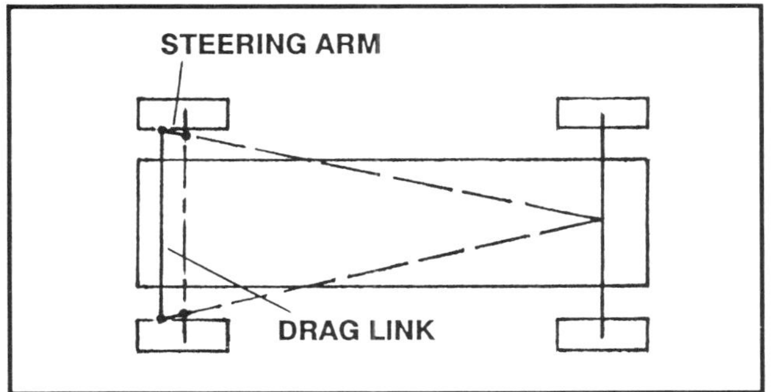

The other reason (a very major reason) is the increase technology in race tire design. The tire contact patch is wider. Racing tire operating slip angles have decreaased quite a bit (which means the direction the two front tires point in a turn is now much more critical). And, today's racing tires are much more sensitive to small changes in slip angle.

Increased Ackerman decreases — or eliminates — cornering understeer. It responds this way in both high speed and low speed cornering, on both asphalt and dirt tracks. Ackerman steers the left front wheel more and thus helps to point the car.

Ackerman steering actually is dynamic front toe-out. It only creates toe-out as the front wheels are steered. Large amounts of toe-out are required to steer a car in to a turn properly, but large amounts of static toe-out are not beneficial to the car. It creates excessive drag on the straightaways and causes darty steering response. However, Ackerman steering will only create toe-out when the car is steered.

How Much Ackerman Steering?

The amount of Ackerman steering a car has is measured in the difference in steering angle between the right front and the left front wheels. Because on oval tracks the right front is the controlling tire, the amount of Ackerman steering is quoted as the amount of steering angle gain at the left front over the right front.

You need to have — for any short track — a 5-degree steering gain at the left front in 18 degrees of right front steering. In other words, if you turn the wheels 18 degrees to the left with the right front tire, the left front should show 23 degrees of steering angle. You achieve this on a front steer car by heating and bending the steering arms outward toward the brake rotors. (This can only be done with a forged steel spindle. Do not attempt it on a stock GM spindle because they are made of a cast material.) Be careful in doing this, and do not quench the hot arms after bending. Allow them to air cool.

On some cars, 5 degrees of Ackerman gain in 18 degrees of steer may not be possible to achieve. In fact, wioth the stock type of Camaro steering and suspension, 3 degrees is about the maximum. Just get as much as is possible into your chassis.

There are other ways of gaining Ackerman other than bending the steering arms. One way is through carefully controlled bump steer. For short track cars (both dirt and asphalt applications), bump steer should be 0 (no steer whatsoever through bump and rebound travel) at the left front and should be 0.040-inch out per inch of upward spindle travel at the right front. An explanation of why this works is in the Bump Steer subchapter in the Chassis Design chapter.

The other way is through repositioning the steering drag link — spacing it back. The drag link can be set back 2 to 2-1/2 inches, or as much as space will allow. This plus changing the angle of the steering arms will quicken the steering ratio at greater steering angles. It will also have an effect on bump steer.

Running a lot of Ackerman in the chassis not only is helpful for turning the car in to the turn, but also for countersteering when a car is loose (both on dirt and asphalt). When you countersteer and have Ackerman present in the steering geometry, the right front will toe out a lot and it will drive the right front out and stabilize the car. However, if there is no Ackerman or the front wheels toe in when countersteered to the right, the front end of the car will be pinned and you will spin out.

You do not want to gain Ackerman by changing the length of one steering arm (such as by shortening the left front). This will open up a whole world full of new problems (this is true on both dirt and asphalt). One of these problems is that with a shorter left front steering arm, the car will gain toe out, as you want, turning left, but it will gain toe IN countersteering back to the right. This is very undesirable.

Checking Your Ackerman

To find out how much Ackerman you have built in to the steering geometry, you need to have steering plates that measure the amount of degrees each wheel turns. You don't have any steering plates? Don't worry — you can build them inexpensively yourself. All you need is two large sheets of cardboard, a protractor for finding angles, a drawing compass, some wire and a little bit of sand.

Cut the cardboard so it is a little larger on all sides than a tire profile setting on the ground. Mark a center point on the cardboard, then use the compass to draw a circle on the cardboard which will be seen when the tire sets down on the center of the cardboard.

Set the cardboard (which is now your steering plate) under each front tire. Sprinkle a little sand on the floor under the cardboard so there is very little friction. You want the cardboard to move with the tire when you steer the wheels. Tape a short piece of wire to the front of the

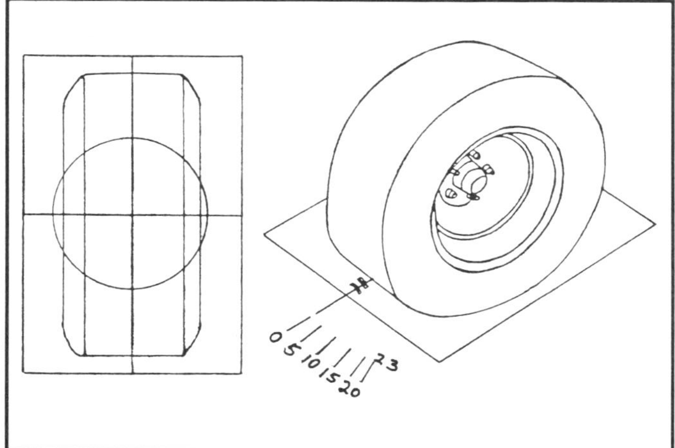

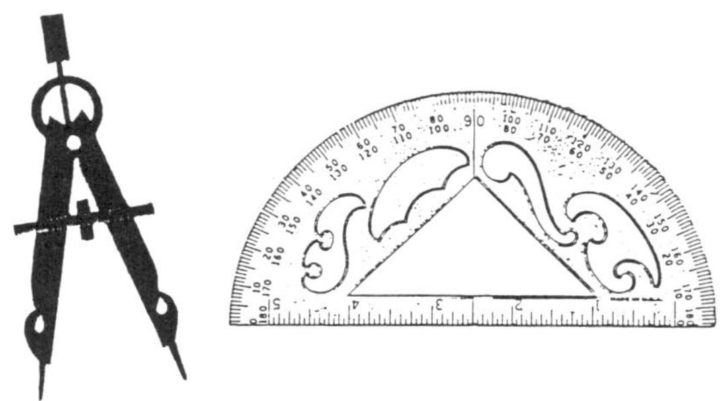

Just to familiarize you with the drawing tools we mention, at left is a compass and at right is a protractor.

cardboard in the center mark of the tire. This will be the pointer.

Use the protractor to pre-mark degree marks on the shop floor so you know how many degrees the tires are being steered. Measure the degrees outward from the center of the circle you drew on the cardboard. Make sure the wire pointer on the cardboard plate lines up with the 0 degree mark you drew on the floor.

With everything in place, steer the car to the left. At, 5 degrees, 10 degrees and 15 degrees of left steer of the right front tire, mark down on a chart the number of degrees the left front is steered.

Bump Steer

Bump steer is the change in steering angle of the front wheels caused by the front suspension moving up or down through its travel. Bump steer causes the introduction of toe (either in or out) into the front wheels when the suspension goes into bump or rebound. Bump steer occurs when the tie rod end follows a path defferent from the path the wheel is following. If you visualize the arc created by the front wheel as it moves up and down through its travel, along with the arc created by the tie rod end, you will see that both arcs must have the same instant center or the tie rod end will move in or out in relation to the wheel, causing a change in toe. If the arc created by the tie rod end and the wheel are parallel, no toe change will be present.

The Vic Irvan bump steer gauge consists of a frame supporting two dial indicators and a hydraulic jack. A graduated plate bolts to the hub. As the jack moves up the plate moves across the dial indicators to show bump steer. A simpler type of bump steer gauge is shown below.

For short track cars (both asphalt and dirt applications), bump steer should be set at 0 at the left front (no steer whatsoever through bump and rebound), and 0.040-inch

There is an easier, though not as accurate, way to measure bump steer than using dial indicators.

Clamp an L-shaped angle metal bar to the face of the brake rotor. Make sure the steering is solidly clamped in the straight ahead position.

Place a square against the L metal at static ride height of the wheel, at both the front and the rear edges of the L metal. Make a mark on the floor. Then move the wheel up one inch in bump travel. Place the square on the same spots at the front and rear of the L metal, and then make another mark on the floor. Now measure the difference in marks at both the front and rear. Use a machinist's ruler for accuracy (read in thousandths of an inch).

The accuracy of this method is in the accuracy of the marks on the floor. They have to be narrow. A broad mark, like made by a piece of chalk, would completely invalidate any meaningful measurement.

out per inch of upward spindle travel at the right front. (See the Bump Steer subchapter in the Chassis Design chapter for an explanation of why this is beneficial.)

Setting Bump Steer

To check and set the bump steer first set the caster, camber and toe-out at the correct static alignment. Clamp the steering wheel tight with the front wheels in the straight ahead position. (Check the centering of the steering by making sure that the inner tie rod pivot points on the drag link are centered in relation to the inner pivot points of the lower A-arms.) Then place the car on jack stands and disconnect the anti-roll bar, springs and shocks. Remove the wheels and bolt a flat plate to the hub. This flat plate should have a hole cut in its center which the hub protrudes through so the wheels studs can bolt through corresponding holes drilled in the flat plate. The plate should have lines scribed or drawn on it at 1/2-inch increments so the amount of travel through bump or rebound of the spindle can be accurately measured.

Two dial indicators that read in thousandths of an inch are attached to a stand which will hold them against the flat plate as the suspension is moved through bump and rebound. Place a jack under the lower A-arm and bring it to normal ride height. Make sure the flat plate is level. Use at least one-inch travel dial indicators. When the dial indicators are placed against the plate at the 0 travel position (normal ride height), compress the indicator needles against the plate slightly and then set both dial indicators to zero. Make sure that both indicators are set equal distance from the plate.

At the right front, you will use the jack to move the suspension through three inches of bump (upward) travel, and one inch of rebound (down) travel. At the left front, you will move through two inches of bump travel and two inches of rebound. Starting at the right front, move the suspension 1/2-inch at a time and make a chart of the bump steer. The difference between the readings on the dial indicators is the measurement of bump steer.

Making suspension changes to correct the bump steer can be a time consuming process. But it is a very important suspension adjustment. Make sure you take the time to do it. And, any time your race car has suffered any damage to the front suspension, steering or frame rails, be sure to do a bump steer check again.

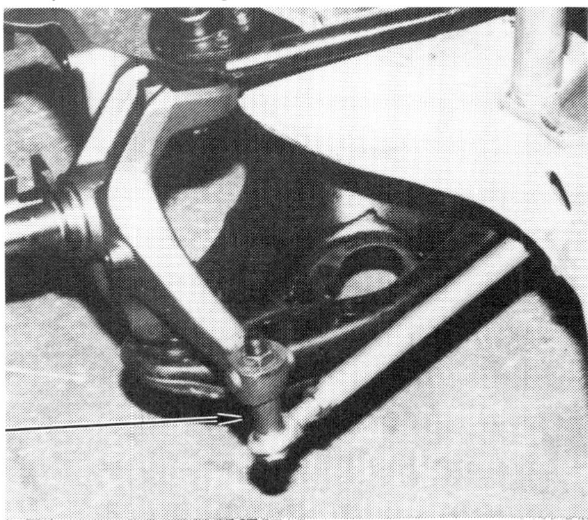

Bump steer can be adjusted by adding spacers between the spindle steering arm and the tie rod attachment to it (see arrow).

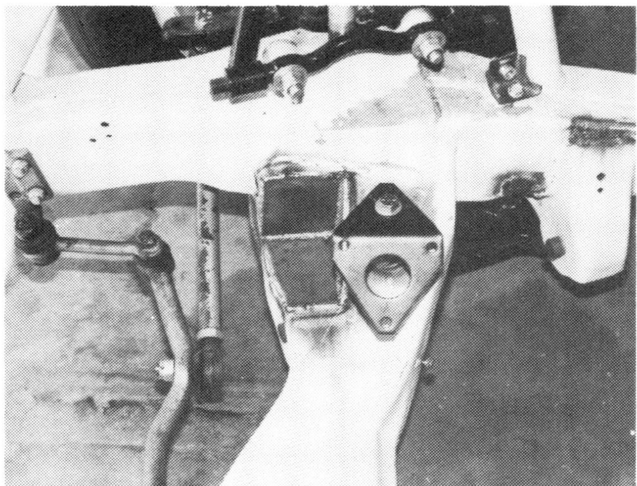

Bump steer can also be adjusted by moving the steering drag link back toward the cross member. Many times, though, there isn't enough room to move it back the required distance.

Spring Rate Vs. Wheel Rate

The spring rate of a spring is the comparative rating of one spring against another in terms of its resistance to a load placed on it. This measurement is expressed in terms of pounds per inch. For example, if 200 pounds were placed on a spring and it compressed one inch, the spring rate is 200 pounds per inch (load divided by inches of compression).

The wheel rate of the spring is the effective rate of the suspension spring at the lower ball joint of the A-arm which compresses the spring (on an independent front suspension). The wheel rate is the spring rate corrected by the mechanical advantage, or leverage. This correction factor is the motion ratio of the linkage squared. The motion ratio, or leverage ratio, is the pivot point to spring center distance (A on the drawing) divided by the total effective length of the A-arm (B in the drawing).

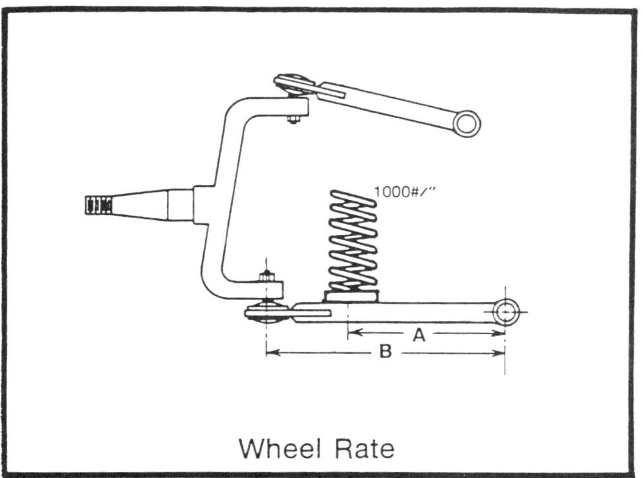

Wheel Rate

Wheel Rate Vs. Wheel Load Rate

Wheel rate, at the front of an independent suspension car, is the measurement in pounds per inch of effective spring rate at the lower ball joint. The effective spring rate is calculated at this point because the lower ball joint is the outer most connecting point of the lever arm (lower A-arm) which creates the leverage ratio or motion ratio that acts upon the spring rate. The wheel rate (or ball joint rate) has been used for years as a common denominator in comparing different springs with the same leverage ratio, or the same spring with different leverage ratios.

Many racers notice that the ball joint is located a distance away from the center of the tire contact patch, and the distance from the ball joint to the tire center varies among different cars. Thus they are concerned about determining the spring rate effect at the center of the tire and not at the ball joint. This rate, called the wheel load rate, takes into consideration the total lever arm extending from the center of the tire contact patch to the inside pivot point. It measures the total effective spring rate at the center of the tire contact patch, or in other words, the wheel loading rate. It is computed by the following formula:

$$K_{WL} = K_S \left(\frac{A}{B}\right)^2 \times \left(\frac{C}{D}\right)^2$$

As you can see, the formula is only a modification of the wheel rate formula. And, if you plug in actual numbers to the wheel load rate formula, you will find that the answer varies VERY slightly from the wheel rate, even with wide differences in wheel offset. And, you can thus see that, for ease of computation and keeping things simple, it is easier and simpler to just use the wheel rate (or ball joint rate).

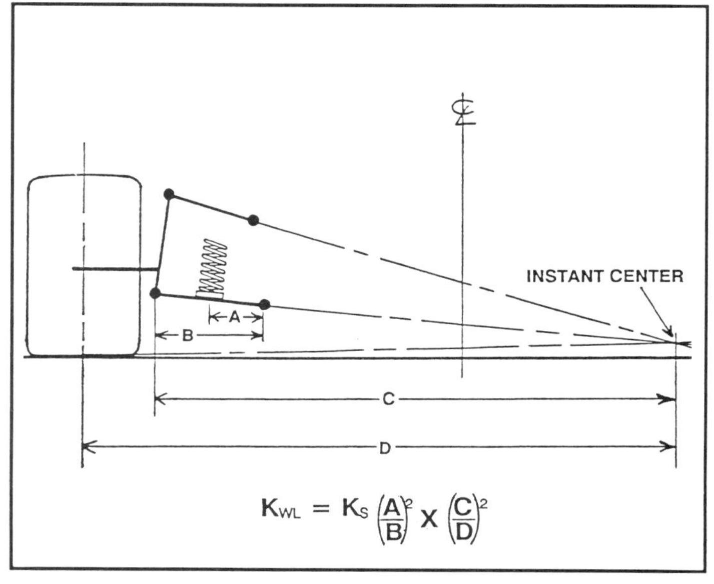

Front Stagger

There is no such thing as front stagger, because the front suspension is independent — the two front tires are not tied together on a common axle. Each tire is independent of each other and a circumference change at one will not affect the other. The only change that will be apparent is that a larger circumference tire is taller than a smaller one, so there will be a corner height difference, which is the same as weight jacking. Most racers, however, refer to this difference as tire stagger. Generally, a left front tire with a circumference one inch smaller than the right front is used on most paved track set-ups. This front tire size difference effects the amount of braking torque applied to each front wheel going into a turn. As brake torques are applied differently, the left front receives more braking torque and pulls the car into the corner.

Anti-Roll Bars

Anti-roll bars (or sway bars) are used to help control body roll and balance out the front-to-rear weight transfer under cornering loads. If only the spring rates were used at the front end to control bump movement, downforce loading and body roll, the spring rates would have to be too stiff for optimum handling and optimum wheel control over bumps. Therefore it becomes necessary to provide a means of controlling body roll separately without interfering with the spring rates and thus the vertical control of the wheels. The anti-roll bar is added to the suspension for the roll control. Of the total front spring rate (both front springs plus anti-roll bar) of a race car, 35 to 50 percent of that total should be from the anti-roll bar. A flat race track would need the most anti-roll bar rate, with a higher banked race track requiring the least anti-roll bar rate percentage.

The best way to balance the front end of your chassis for oversteer/understeer is to always keep your basic spring rates at the left front and right front, and change the anti-roll bar rate to change the roll couple balance. You should always have at least two anti-roll bar rates with you at the race track, and better, have three available. The wheel rates of the bars should be in the range of 225, 325 and 425.

Fortunately for the Pro Stock racer, there are a wide variety of production type anti-roll bars available in wrecking yards that adapt very well under the front of the Camaro frame. An example is the following three parts found on various models of Firebirds (each bar is 34 inches long, with 11-1/2 inch long arms):

Part Number (GM)	O.D.	Spring Rate
3984557	1.000	248 #/"
3975523	1.125	396 #/"
3986480	1.250	605 #/"

There is a wide variety of production type anti-roll bars out there for you to choose from. You are only limited by your research time. Try looking at vehicles you would suspect would logically utilize a stiffer anti-roll bar: high performance options of passenger cars (such as the Firebird T/A or Camaro Z-28), 1/2-ton and 3/4-ton trucks and vans, the Chevy Blazer, etc. When you go to the wrecking yard, take along with you a caliper to measure diameter and a tape measure to measure bar length and arm length. You can calculate the spring rate of any solid anti-roll bar using the formula found in "Advanced Race Car Suspension Development" (order number S105), published by Steve Smith Autosports. Going into that formula here is out of the scope of this book.

Another suggestion, if you cannot find a suitable production bar, is to fabricate your own. Obtain an appropriate length of 4340 alloy steel round stock in the desired outside diameter, and bend it to the length and arm length you require. Then have it heat treated to Rockwell 35 on the C scale.

Use 5/8-inch spherical rod bearings (one male, one female, per side) to mount the anti-roll bar to the A-arms on your race car.

Scrub Radius

The scrub radius is the distance from the upper and lower ball joint projection line (or steering axis line), as it meets the ground, to the center of the adjoining tire contact patch. The scrub radius creates a leverage effect on the A-arms and spring. Wheel offset amount and the steering axis inclination angle of the spindle both have a bearing on the width of the scrub radius.

With a scrub radius, when you turn left, it lengthens the right side wheelbase, which tends to loosen the car. When you countersteer, it shortens the right side wheelbase and it puts understeer in the car. This is a very stabilizing effect to the control of the car and in feedback to the driver.

The opposite of scrub radius is center point steering. That is where the steering axis inclination intersects the center line of the tire contact patch. With this situation the driver has no seat-of-the-pants steering feel or feedback at all.

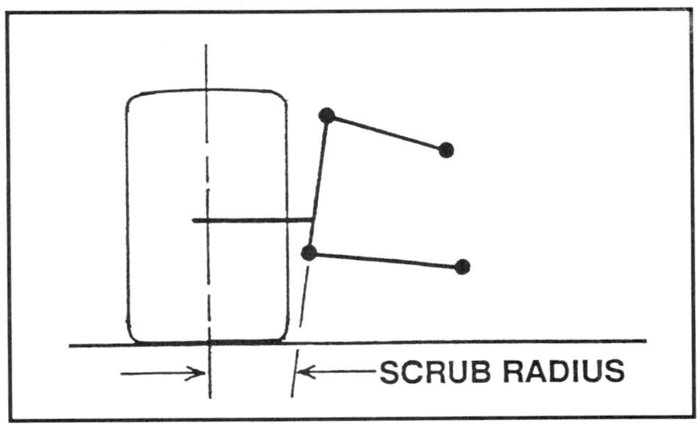

This is the "good" power steering box, which is found in the 1982 through 1986 Camaro Z-28 and Firebird Trans Am.

The driver will complain about the front end moving around and he can't feel what the car is doing. Many racers who have never done a scale drawing of their chassis have this problem, and have made other changes (spring rates, excessive positive caster, etc.) to try to correct the situation. Most race cars, with 8 to 11 inches of tire tread face on the ground, will have a 5 to 5.5-inch scrub radius. This is the outer bounds of acceptibility. About the very best that can be achieved with a Camaro type of suspension is 4 inches, but this requires a highly inset wheel.

Power Steering

Many racers argue that power steering on a race car lessens the driver feel and control of the car. But the type of advanced power steering units developed today, the driver loses very little feel through the tires, and it actually helps the driver retain better steering control over his car. Today's high performance GM boxes are a significant improvement over earlier "slushy" feeling power steering boxes. And with a 3,000 to 3,200-pound Pro Stock car, the extra steering assist is a big help. This is especially true when using the Camaro-based A-arm linkage with its inward-angled swing plane because this causes a large amount of positive caster gain under bump travel, which greatly increases steering effort.

The recommended power steering box to use is found in the 1982 through 1986 Camaro Z-28 and in the 1982 through 1986 Firebird Trans Am. The GM part number for this box is 7839897, and the Hollander interchange number is 1271. This is a Saginaw self-contained or integral power steering box. It is a constant ratio 12.7 to 1 gear ratio box.

Stay with the steering box we recommend, or you might run into some problems. Although a lot of Saginaw steering boxes look alike, they are not all the same thing inside. Some boxes feature a constant steering ratio, while some have a variable ratio. The variable ratio is very undesirable for racing applications because the steering wheel input at different points will be different from box work output. The other problem is that, even though they all look the same on the outside, the integral style of steering box actually has been available in three different models. These are basically a light duty, medium duty and heavy duty box. The difference is the strength of the internal components. The "good" box has three physical identification items that sets it apart from the other boxes: 1) it has a square side cover, 2) it has a 1.25-inch diameter pitman shaft (which requires a correspondingly larger pitman arm), and 3) the large end bore inside diameter measures either 3.875 or 4.25 inches.

The required pitman arm application is from the 1976 through 1981 Firebird Trans Am.

All years of 1970 and later Camaro and Firebird have the same steering box bolt pattern bolting to the frame, so there is no need to worry about box interchange. Some models of the heavy duty steering box will be found with a fourth mounting lug on the housing. This was just added to give added mounting strength in some chassis applications. Don't worry about it if you end up with one of these boxes — just use the standard three-bolt pattern.

Power Steering Pumps

A stock GM pump cannot be used because a high performance application requires a greater pump capacity in both volume and pressure. Modified pumps for racing are available from Wilwood, Lee and Sweet. The new Wilwood unit features mounting brackets that easily line right up. The right power steering pump can be reworked, but we will not outline any procedures here because doing it incorrectly can adversely affect the steering operation of your car — and we certainly don't want that. Your best bet is to acquire a high performance pump from a reliable manufacturer.

For power steering fluid lines, use only steel braided teflon core -6 (dash six) size lines. Be careful — output pressures can exceed 2,000 PSI.

Use only genuine GM power steering fluid. Do not use automatic transmission fluid in your power steering pump.

Steering Universal Joints

Steering universal joints are a very critical link in your steering system, and should be chosen and installed very carefully.

Although universal joints are designed to operate at an angle, increasing the operating angle past 30 degrees significantly increases the stresses imposed on the joint. Do not do it. If your steering application requires operating a u-joint at close to 30 degrees, you should redesign the steering to employ two universal joints.

The connection of the joint to the steering shaft is very critical. There are three different ways that racers use: 1) welding, 2) pinning, and 3) splined shaft and u-joint coupling. Of the three methods, a splined shaft into a splined yoke is definitely the very best, strongest and safest way to make a connection.

Welding can be a very dangerous situation. Not only does the driver's safety and health depend on the quality and integrity of the weld, but welding heat (even very low temperatures) can cause serious damage to the joint. Welding can distort the housing, freeze up the pins on a needle bearing joint, and even cause the pin to inadvertantly weld to the yoke. Another important consideration is that welding the shaft to the joint makes it a very permanent connection, something that is not very desirable in a race car.

Pinning the yoke to the shaft is done by pressing in two roll pins or shear pins at right angles to each other a distance apart on the steering shaft. The problem with this is that the holes through the steering shaft severely weaken the connection area of the shaft. And, again, ease of assembly and disassembly is critical here as the pins have to be driven out to disassemble the unit. This is not an ideal situation on a race car where time and ease of assembly is important.

The splined shaft and yoke is the best way to effect a connection. It is strong, it is safe, and it is easy to assemble and disassemble. After the steering system is laid out and the position of all components is determined, the joint is held firmly to the shaft with a set screw. File or grind a flat spot on the steering shaft where the set screw will set snugly to the shaft. Use Loc-Tite or a similar product on the set screw when installing it to prevent it from backing out. And, periodically, check the tightness of the set screw. Make sure that at least 3/16-inch of the set screw threads are engaged on a hardened steel shaft, or at least 5/16-inch of threads on an unhardened steel shaft so the threads in the yoke don't rip out. Make sure there is a full engagement between the spline shaft and the splines of the yoke.

An easy type of steering shaft support. However, note that this one is too confining around the lock nut, making it very difficult to put a wrench on it. Use a piece of channel or cut a piece of rectangular tubing.

A simple piece of angle iron will make a good steering shaft mount.

Be sure to use spherical rod end bearings as a support for the steering shaft. The bearings must be properly attached to the chassis structure. AFCO offers a special spherical rod end bearing designed especially for us as a steering shaft support bearing. The bore of the bearing is is 0.007-inch oversized so the rod end slides right over 3/4-inch steering shaft material — no need to grind out a standard rod end bore. The part number is 10400 (right hand thread).

If two universal joints are used in the steering layout, make sure the forks of each are installed in line with each other. And, make sure the steering layout allows for collapse of the steering shaft in case of an impact away from the driver, and that it is not pushed into (or through) him.

There are several different brands of universal joints to choose from. One company's product that we can recommend is Borgeson universal joints. They manufacture products specifically for racing and high performance use, they are torque-tested to over 7,500 inch-pounds, and have less than 0.001-inch backlash.

Steering U-joints should be of the very highest quality for racing use. Use joints that have a solid block between the two halves, not a split block.

Chapter 5

Rear Suspension

On any type of track, the key to performance is getting good bite and acceleration from both rear tires. This involves an interaction of weight transfer, induced downforce loading of the tires, and the proper compliance at the tire contact patch. The relationship between the tire contact patch and the track surface can be a very fragile one when the tires are subjected to both lateral acceleration and forward thrust loadings. This relationship can easily be destroyed by a harsh torque reaction or a violent suspension movement. The spring rates and suspension attachment linkages must be carefully designed to cushion these harsh and violent reactions, and preserve the fragile contact patch/track relationship.

Let's take a survey of the various rear suspension systems being used in stock cars today, and try to determine how each affects this relationship.

The Three-Link System

The three-point rear suspension linkage has the most wide-spread useage for all stock cars employing coil springs at the rear, especially on paved tracks. The traditional three-point linkage includes two lower links about 24 inches in length, attached one at each outer end of the rear end housing from a bracket on the housing straight forward to a bracket on the chassis. The third link is attached at the center on top of the rear end housing, running forward to a bracket on the chassis. The forward mounting height of this upper link can be adjusted for the amount of anti-squat in the rear suspension. The three-point rear linkage is very easy to work with and is highly adjustable so that it can be adapted to almost any type of racing condition.

Some chassis manufacturers are designing a three-point system with the lower right side trailing link considerably longer than the left side link. The reason for this is simply to eliminate roll steer during cornering. The longer the trailing arm, the flatter the arc the end of it scribes as it moves in bump and rebound. With a trailing arm which measures 40 to 50 inches, the arc is so flat in four inches of travel that there is hardly any noticeable amount of wheel displacement.

The typical three-link rear suspension system for paved track application.

The geometry — angles — of a three-point link system, projected forward to a common point, form an instant center, and thereby actually behaves like a torque arm (refer to the subchapter on torque arms). Because of this, be very careful in the layout of the three-point rear linkages so that this instant center doesn't have a lot of movement during rear suspension travel. Having very steep angles of the upper and lower arms will create a short instant center where it moves around. If the suspension travel causes instant center movement, the rear wheels and suspension arms will be changing angles quite rapidly, making for a very unstable feeling for the driver.

Another important point to keep in mind with a three-point suspension is that a torque absorbing link in the upper third link position can make a big difference in handling, especially when a "track tire" is required. With a hard tire compound, the torque reaction will break the rear tires loose much quicker and easier than with tires that are more forgiving. The torque absorbing upper link can help cushion the torque reaction at the tire contact patches.

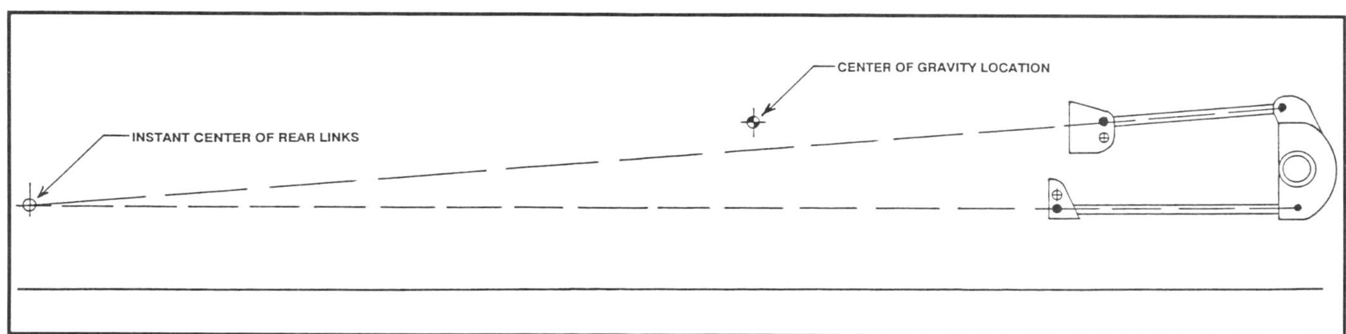

Above, this type of link layout yields a very long instant center and a very stable feeling car under both acceleration and deceleration. Steep angles of the upper and lower links, such as shown below, makes for a very unstable feel.

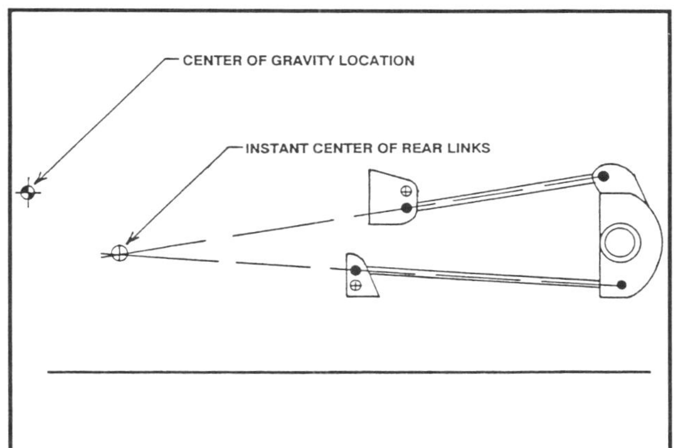

A torque absorbing upper link will help cushion the torque reaction at the tire contact patches.

The angle of the upper link in a three-point system has a very important influence on the performance of the chassis under braking and acceleration. If the upper link is mounted at a downhill angle toward the front, this will promote more anti-squat under acceleration. This means the rearward pitching moment of the car under aceleration is reacted by the mechanical leverage of the suspension arms and thus causes additional downforce to be

Above, trailing arm links should be mounted in double shear, or straddle mount, brackets. The material is only 3/16-inch thick for each bracket, but the double brackets add considerable strength to the mount. Note that there is distance on either side of the rod end to allow for lateral movement of the arm. Also, spacers are used on either side of the rod end to keep it aligned and to allow for maximum misalignment travel of the rod ball.

A single shear mounting such as this offers several disadvantages. Even though the bracket is thicker and heavier than the 2 double shear brackets above, the double shear mounting is still stronger. And, the single shear mount doesn't offer as much misalignment room.

Some type of flexible retaining strap should be used to anchor the rear end housing to the chassis to prevent the spring from falling out under maximum droop.

Proper coil spring mounting position for a paved track car. Note the right side chassis mount for the Watt's linkage.

Note the placement of the spring and the shock absorber, both as close as possible to the wheel. Also note the left side mounting bracket for the Watt's linkage.

placed on the rear tires. However, the big drawback of this is the opposite and equal reaction under braking. A large amount of anti squat (downhill angle of the upper link) will cause rear end lightness and rear wheel hop under braking. The more downhill angle of the upper link, the more severe the problem will be.

An uphill angle of the upper link will promote good firm traction of the rear wheels under braking. But it has a big drawback under acceleration — it promotes pro-squat, which lifts the rear wheels up under acceleration and diminishes traction.

The answer is a compromise. The upper link should angled downhill at a seven-degree angle. This promotes good traction without the associated problems of rear end lightness and wheel hop.

The rear mount of each lower trailing link should be located within 8 to 12 inches of the brake rotor rear face. If the mounts are too far inboard to the center of the car, the leverage forces are going to be real hard on the rod ends' life.

Be sure to check the rod end bearings for wear or freeze-up. These problems can create difficult-to-diagnose handling problems. Check rod end freedom weekly and keep them cleaned and lubricated.

Leaf Springs

The leaf spring has long been the traditional spring mechanism for stock cars on dirt tracks. This is because the leaf spring twists and deforms, making it a very forgiving type of spring. Because so much of the acceleration torque reactions are cushioned by the deformation of the leaf spring, it is very difficult to destroy the tire-to-track contact patch (provided, of course, that the driver uses proper driving techniques).

Another reason leaf springs have traditionally been preferred for dirt is that well-designed leaf springs help to

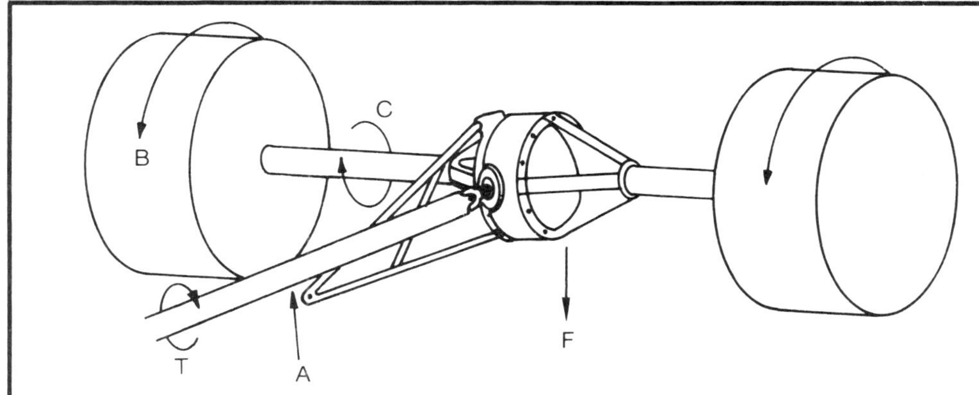

Engine torque input to the rear end (T) creates a driving force at the wheels (B) and an opposite reaction at the rear end housing (C). This reaction forces the torque arm up at A where it is resisted by the weight of the chassis and produces an equal downforce reaction (F) on the rear end housing.

positively control the location of the axle. Many early attempts at properly controlling rear axle location through bump and rebound with 2, 3 or 4-point linkages in coil-sprung cars were very badly done. This created an impression in many racer's minds that the coil spring suspension was responsible for the resultant poor handling characteristics on dirt. No so. The real problem was that of car builders arbitrarily attaching suspension linkages without analyzing the travel path they created for the rear axle.

In today's more sophisticated brand of stock car racing, however, the leaf spring system's failings are being understood, and leaf spring suspensions are being phased out in favor of coil spring systems. Note that we are not saying here that leaf springs are "bad." Leaf spring suspensions are a viable alternative on dirt for the novice low bucks racer, but they are not the ultimate in sophistication for the modern dirt track car.

Leaf springs have many inherent unfavorable characteristics. First and foremost, they are very heavy, thus increasing the unsprung weight. And they are bulky — they are harder to fit in many places.

Leaf springs are difficult to use for fine-tuning the suspension. The average racer cannot afford to carry with him 7 to 10 sets of leaf springs, each 50 pounds per inch different in rate. He can do this with coil springs. Plus, changing a leaf spring at the track is much more difficult and time consuming than changing a coil spring.

The other drawback of using leaf springs is that it is difficult to change the rear roll center height with them. It is built in at the attaching point of the spring to the axle housing. It can be done by inserting lowering blocks between the housing and the spring. This also will change the cross weight in the car. With a coil-sprung car, however, the roll center is determined by the height of the lateral locating linkage — such as the Panhard bar — and can be varied a great deal.

It is our position that a coil-sprung car can easily be made to work better than a leaf spring car. There are, however, thousands of leaf-sprung cars out there, and they can be competitive with coil-sprung cars.

Torque Arm Suspensions

The torque arm is a traction device which is solidly mounted to the rear end housing and runs forward to a pivot point which attaches the arm to the chassis. The torque arm uses force created by the rear end. It rotates up about its axis under acceleration. The torque arm, being soildly attached to the rear end, harnesses this force and applies it upward against the weight of the chassis and body. The opposite and equal reaction to this force pushing up against the chassis pushes down on the rear end and tires, creating more down force on the tires, and thus creating more traction.

The amount of traction a tire will develop is directly related to the amount of load on the tire. When the load is increased, such as with the torque arm, the available traction of the tire increases.

The actual force applied on the torque arm, and back to the chassis, is the engine torque (in foot/pounds) multiplied by the final drive ratio, all divided by the length of the torque arm (in feet). The answer is expressed in pounds of force. The formula would be:

$$F = \frac{T \times R}{L}$$

where **F** is force downward applied in the rear end housing, **L** is the length of the torque arm, **T** is the engine torque and **R** is the final drive ratio. Let's look at an example:

$$F = \frac{500 \text{ ft/lbs} \times 5.60}{3}$$

$$F = 933 \text{ pounds of force}$$

This is a significant amount of force being applied downward on the rear end housing to encourage tire bite. And, you can see from the formula the amount of force being reacted at the rear end is directly proportional to the amount of torque input from the engine. If the torque is

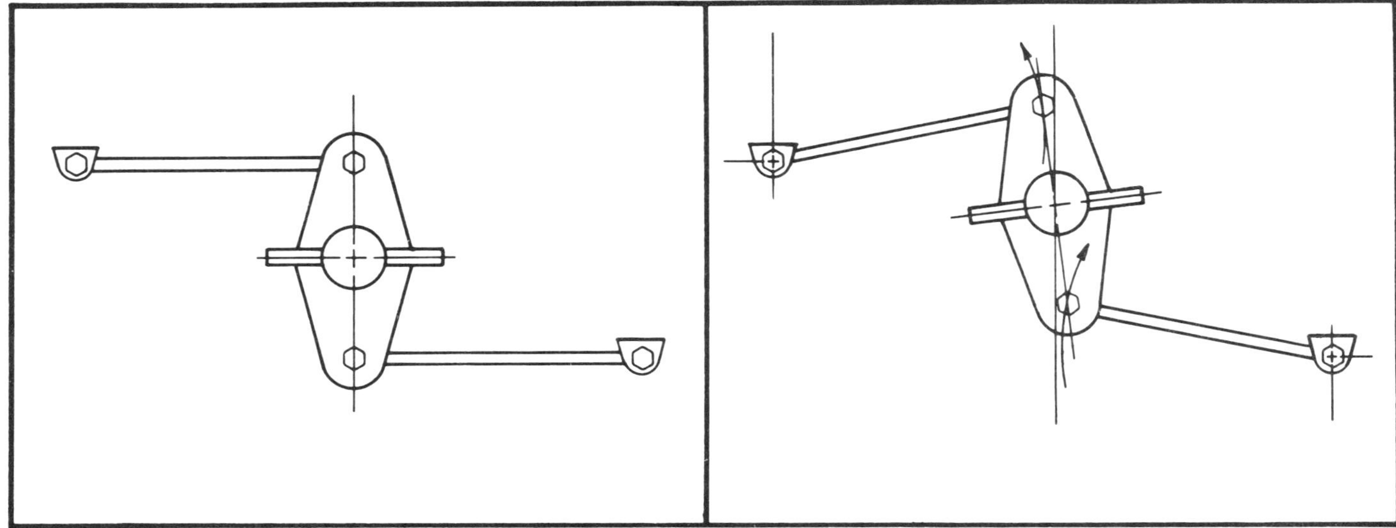

This demonstrates how a Watt's linkage arrangment keeps the axle perfectly centered through bump and rebound travel. This is true for a Watt's link used to center a birdcage or as a lateral locating device for the rear end housing.

A Watt's link used to control a birdcage assembly. It works with a coil-over, but there isn't enough room with a separate shock and spring.

fed violently all at once (driver mashes accelerator), the opposite and equal reaction is going to be violent. This will upset the chassis and handling with a tremendous shock loading at the tire contact patches.

Torque Absorbing Suspension Systems

This is the era of shock control at the tire contact patch. Racers — on dirt and asphalt — have discovered that a race car can get a better bite and get hooked up better upon acceleration off a turn if the torque application at the rear tires is cushioned.

The evolution of torque absorbing systems has gone from leaf springs (which cushion initial shock loading with spring wrap up) to softer leaf springs coupled with a torque absorbing shock absorber, and then to the fifth coil-over torque arm system which mounts the rear end housing in birdcages (or floater brackets) on each side.

The birdcage or floater bracket, mounted on each side of the rear end, is the real key to making a torque arm system operate properly. The birdcages are free to pivot about the rear end housing, being restrained only by the attaching linkage arms to the chassis. This leaves the rear end housing free to pivot about its axis and apply loads through the torque arm without any restraints of other attaching linkages to the housing to bind it or interfere. The attaching linkage arms on the birdcages control or restrain the fore/aft movement of the housing.

There are two different types of linkage arrangements which are used to locate the birdcages — the Watt's linkage and the wishbone linkage.

The Watt's Linkage Trailing Arm

The Watt's linkage employs two equal length torque arms, one mounted to the bottom of the bridcage and running forward, the other mounted to the top of the birdcage and running rearward. As long as the two mounts on the birdcage are spaced equally (such as 7 inches from center of housing to center of rod end mount of trailing arm), the rear end housing will move straight up and down in body roll, creating no roll steer. This is because the arc of each trailing arm is equal, but in the opposite direction of each

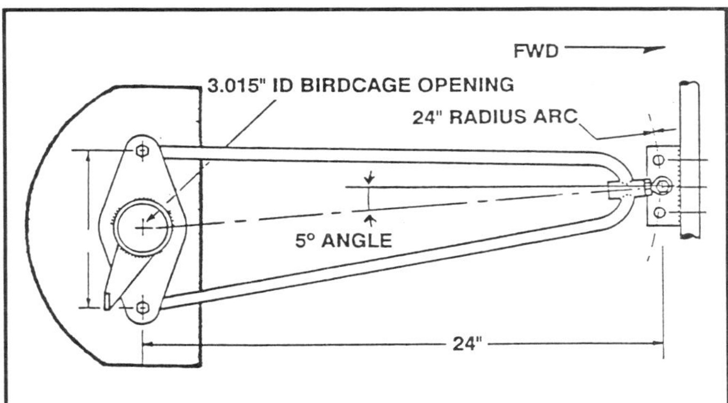

The wishbone trailing arm encompasses all the positive aspects of the Watt's link attachment linkage, but substantially simplifies the linkage design.

This wishbone link utilizes a rubber bushing in it to cushion the input forces to the tire contact patch.

other, so the result is only a pivoting of the birdcage bracket about the housing. The only drawback to the Watt's linkage arrangement is a lack of space on the chassis for the rearward facing trailing arm. This can be solved by using the wishbone link.

The Wishbone Radius Rod Trailing Arm

The wishbone trailing arm link, designed by George Gillespie of PRO Shocks, preserves the same geometry as the Watt's linkage, but simplifies the mounting and adjustment. The wishbone link really represents a refinement in simplicity for the birdcage attachment system. It is simpler and more compact in design than the longitudinally-mounted Watt's link.

The wishbone link uses a trailing arm attachment at both the top and bottom of each birdcage bracket, but the arms converge at their forward end and attach to a single spherical rod end bearing mount, which then attaches to a chassis-mounted bracket. The chassis bracket has a provision for three vertical mounting positions, each spaced two inches on center. The center mounting position on each side is the desired placement for almost all competition situations. It provides for a 5-degree uphill mounting of the links. However, on a dry, slick track, the mounting position should be moved to the upper mounting hole on each bracket to increase rear bite. On a very tacky dirt track, the forward mounting position of the left side bracket should be lowered to the lowest position (with the right side kept in the center position) to unload some of the traction ability of the left rear tire (it makes the car drive more off the right rear).

Note on the drawing that the three holes on the chassis attaching bracket are not drilled straight up and down, but rather on an arc. The 24-inch radius arc provides for zero wheelbase length change when the arm is mounted in any of the three mounting holes.

Why does a slightly uphill angling of the wishbone rods improve traction? If the front of the rods point upward, they will try to lift the sprung mass of the car under acceleration as the rods propel the sprung mass forward. For every pound of forward driving and lifting force, there is an equal and opposite force reaction, with the sprung mass applying that force back down through the links to the rear end housing and to the tires, creating more traction by pushing the tires into the ground.

The wishbone link system produces little, if any, roll steer during cornering as long as body roll is consistant side for side (the right side goes down an inch and the left side comes up an inch). If both sides are the same length, there will be no roll steer because one side pulls the axle forward the same amount as the other side pulls the axle forward in equal but opposite arcs.

This system is very simple to adjust and to square the rear end. You can simply run the rod end on the link in or out to adjust the length.

The wishbone link is a much preferable system to a forward-pointing four link system. This is because the forward pushing energy from the tires and rear end to the car is concentrated at two parallel spots on the chassis with the wishbone link. With the four link system the arms are split in their mounting planes, and this splits the energy fed into the chassis from the rear tires, sending it out in different directions.

The Fifth Coil-Over Torque Arm System

To shelter and protect the rear tires from violent shock reaction, a torque absorbing device is placed between the end of the torque arm and the chassis. This coil-over unit placed into the torque arm system allows the downforce to be more gently applied at the rear tire contact patches as the throttle is applied.

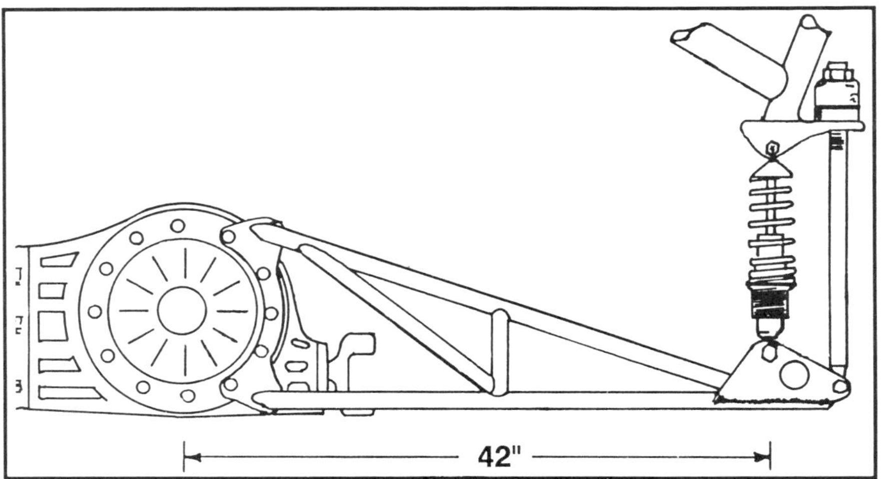

The fifth coil-over torque arm, utilizing a rubber-mounted snubber at the front for braking torque reaction.

This suspension also employs a snubber, or a travel limiter, which limits downward travel of the front of the torque arm under braking torque. (This is the case when the brake calipers are rigidly attached to the rear end housing.) There is as much force pushing down on the front of the torque arm during braking — and consequently up on the rear end housing (thus unloading the rear tires) — as there is in the opposite direction during acceleration. The snubber limits this so the rear wheels don't become totally unloaded. But, the best way to combat this braking reaction is to mount the rear end housing in birdcages, or floater brackets, and then mount the brakes on their own birdcages. Because the caliper birdcages float independently about the rear end housing, the braking torque reaction is fed into the chassis through their own attaching control arms, and not the rear end housing. Thus the torque arm handles almost no brake reaction torque.

The brake calipers must never be attached to the birdcages which position the rear end housing. If they are attached to these floaters, the calipers will lock up the attaching linkages under braking. This will cause severe, or terminal, oversteer. The brake calipers should either be mounted separately on their own birdcages or directly onto the rear end housing. If the calipers are mounted to the rear end, braking torque can be absorbed and controlled by another device. This device could be a 90/10 shock absorber mounted on top of the housing pointing forward and uphill, or a slider bar link in the third link position, or a sixth coil on the end of the torque arm.

An important point to note when using a fifth coil-over system with floater brackets is that the fifth coil-over on the end of the torque arm is used to set the pinion angle of the rear end. Adjusting the overall coil-over unit length adjusts this. The desired static setting is a 2-degree negative pinion angle. The car will experience a pinion angle travel of 4.5 to 5 degrees positive under acceleration, which means the angularity of the driveshaft will be a maximum of 2.5 to 3 degrees positive under acceleration. Under braking, the pinion angularity should travel to a maximum of 4.5 degrees negative, which means a movement of 2.5 degrees past the static setting.

Choosing The Fifth Coil-Over Spring Rate

Through experience it has been found that the fifth coil-over suspension system works best when the fifth coil-over compresses between 2.5 and 3 inches. The

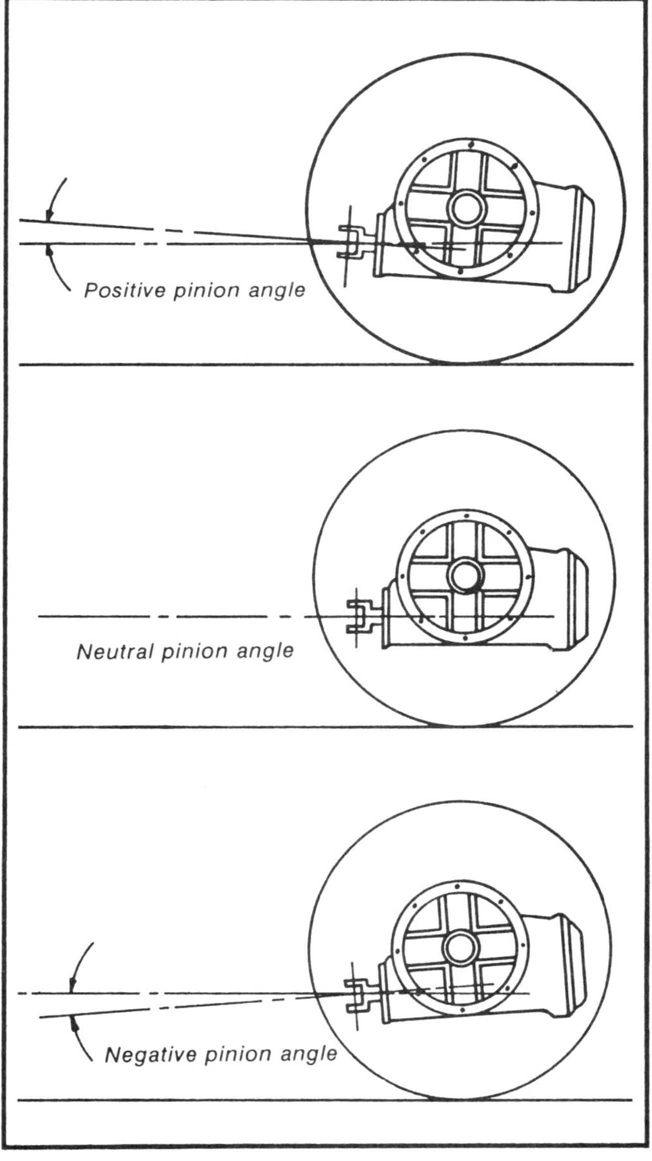

following formula can be used to calculate a starting ballpark spring rate to achieve the spring compression desired:

$$SR = \frac{ET \times F}{C \times TL}$$

where **C** is the desired spring compression, **SR** is the spring rate (which is the unknown facotr we solve for in the equation), **TL** is the torque arm length (expressed in feet), **ET** is the engine torque (expressed in foot/pounds — the torque figure is the torque at average operating RPM range), and **F** is the final gear ratio.

Let's look at an example of the formula. Choose the low range of spring compression, so the value for **C** would be 2.5. The torque arm length is 42 inches, which is 3.5 feet, so **TL** is 3.5. The engine torque at average RPM range is 455 foot/pounds, so **ET** is 455. The final gear ratio is 6.30, so **F** equals 6.30. We are solving for **SR**, so that is our unknown. To solve:

$$SR = \frac{455 \times 6.30}{3.5 \times 2.5}$$

$$SR = \frac{2{,}866.5}{8.75}$$

$$SR = 327.6$$

The closest spring rate application (looking at the PRO Shock catalog) is either a 300 or 350 pound per inch spring (this would be an 8-inch tall spring). It is usually better to opt for the stiffer spring rate.

The formula can usually get you in the ballpark. Then, test the car on the track and look at the amount of compression travel it is getting. The optimum travel is between 2.5 and 3 inches. Anything less than 2.5 inches means you need a softer spring rate. More than 3 inches of travel indicates you need a stiffer spring rate.

Different driving styles will also dictate different spring rates. A driver who accelerates by trying to mash his foot through the floorboard is going to require a stiffer spring — maybe as high as 600 pounds per inch. On the other hand, a smoother driver who pushes the gas pedal like it has an egg under it will require a softer spring rate — maybe in the 250 to 300 pounds per inch range.

Generally, a driver who finds a spring rate that gets the car feel to what is comfortable for him on a tacky track will stay with that spring and not make changes. He will adjust his driving style for track surface changes. However, on a real dry/slick track, he may want to use a slightly softer spring rate.

Many of the problems of choosing the correct fifth coil-over spring rate for different applications has been solved now with the introduction of a progressive rate coil-over spring by PRO Shocks. In the first inch of compression, it has a rate of 175 #/". In the second inch, it is 250 #/". In the third inch, it is 350 #/". And in the fourth inch of compression, it is a 500#/" rate. It very favorably emulates the action of a fiberglass or carbon fiber flexible torque arm.

Fifth Coil-Over Torque Arm Length

The basic set-up for a 108-inch wheelbase car with 52 to 53 percent rear weight on an average 1/2-mile dirt track would be a 42-inch long torque arm.

If the torque arm is too short, it will raise the rear of the car under acceleration and buzz the rear tires. If the torque arm is too long, it will pick up too much of the weight of the vehicle. What happens here is it will lift the front of the car on acceleration and the steering control will get real light.

The Carbon Fiber Torque Arm

The carbon fiber torque arm, manufactured by PRE (Professional Racer's Emporium) is a torque arm made of supercomposite materials, carbon and Kevlar. It is a hollow round tube, 1.25 inches O.D., and 54 inches long.

The carbon fiber torque arm is a very effective system, and in fact, may even be a little more effective than a rigid torque arm with a coil-over unit. And, if track rules prohibit a torque arm with a coil-over device, the carbon fiber torque arm is the way to go.

The carbon fiber torque arm is a pure torque arm in the sense that it is rigidly connected to the rear end housing and is mounted to the chassis in a rigid bracket. This chassis-mounted bracket is actually a slider bracket which moves along a square tubing track attached to the chassis. The slider bracket adjusts the effective length of the torque arm, which adjusts the spring rate of the torque arm.

PRE utilizes the carbon fiber torque arm on the cars they build with two different types of suspension arrangements, the leaf spring and a 3-point system. Using it with leaf springs, the leaves are mounted to the housing on floater brackets to get all of the acceleration thrust reaction into the torque arm. With the 3-link system, the two lower trailing arms are solidly mounted to the housing like in a regular 3-point system, and the housing is allowed to pivot about the balls of the rear spherical rod end bearings on the trailing arms.

With both the leaf spring and 3-link systems, a third upper link is mounted above the center of the rear end housing to provide a positive stop on rebound during braking. This link is a telescoping rod with a spring-rated polyurethane bushing and only comes into play under braking. It is free to expand under acceleration so all the torque is fed into the torque arm. Under braking, it has a

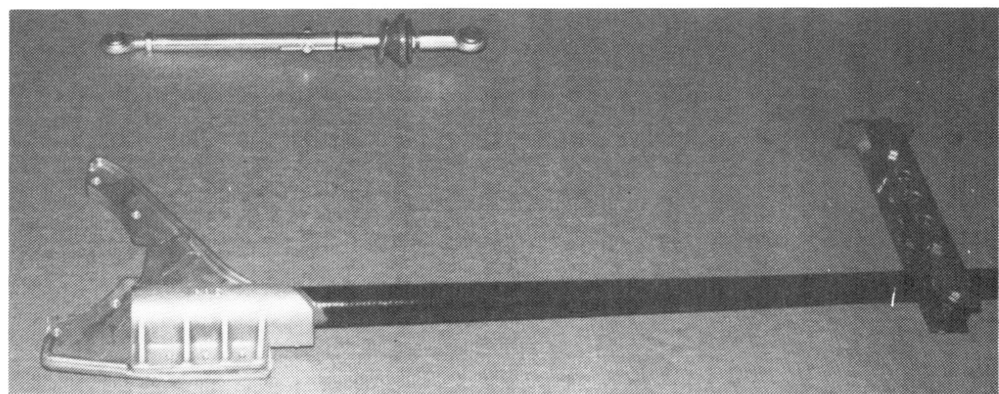

Right, PRE's carbon fiber torque arm system.

When the carbon fiber torque arm is used with leaf springs, the leafs are mounted on floater brackets.

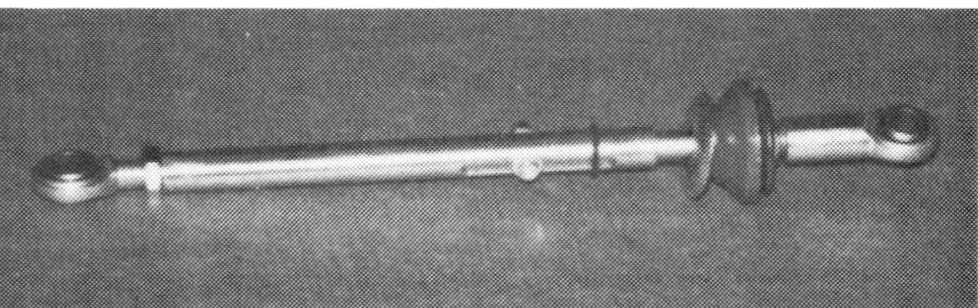

The telescoping upper link is an important element for adjusting the effectiveness of the torque arm system. Below is an illustration of how forces are applied to the tires under braking.

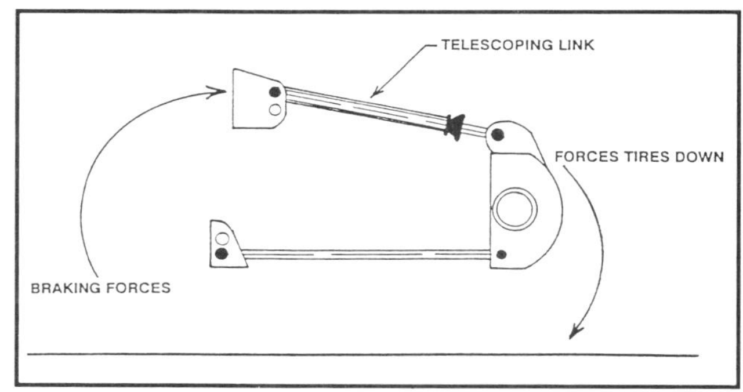

cushioning effect initially, and the link is mounted at an uphill angle (toward the front) so the braking forces push the rear end down under braking.

The top slider bar is a critical element in adjusting the effective length of the torque arm for its optimum effectiveness. A slot on the telescoping link pushes an O-ring back as it moves. When the torque arm is working properly it should rotate the pinion up 4 degrees. This translates to 1.75 inches of movement of the O-ring on the telescoping link. If the O-ring pushes too far, shorten the effective length of the torque arm. If it does not move back 1.75 inches, lengthen the effective length of the torque arm. The effective length of the torque arm adjusts between 36 and 48 inches long. The shorter length yields the maximum amount of traction on a wet track. As the track dries out, move the roller assembly forward for desired traction. Just keep adjusting it to get the 4 degrees of pinion angle change.

For dirt track use, the carbon fiber torque arm should have two mounting positions in the car — one offset to the left of the driveshaft and one offset to the right. This would require two mounts on the rear end housing and two chassis-mounted tracks. For wet, heavy dirt tracks, you want the car to drive off the right rear wheel, so the torque arm should be mounted on the right side to give more leverage to that wheel. For a dry, slick track, you want the car to drive off the left rear wheel, so the torque arm should be mounted on the left side.

If the carbon fiber torque arm is used on a paved track application, the track and arm should be offset to the left of the driveshaft for maximum leverage on the left rear wheel.

The mounting height of the front of the torque arm sets the pinion angle of the rear end. The torque arm should be installed with the arm parallel to the frame rails when the rear end is set at static running height and the pinion is at a six-degree negative angle.

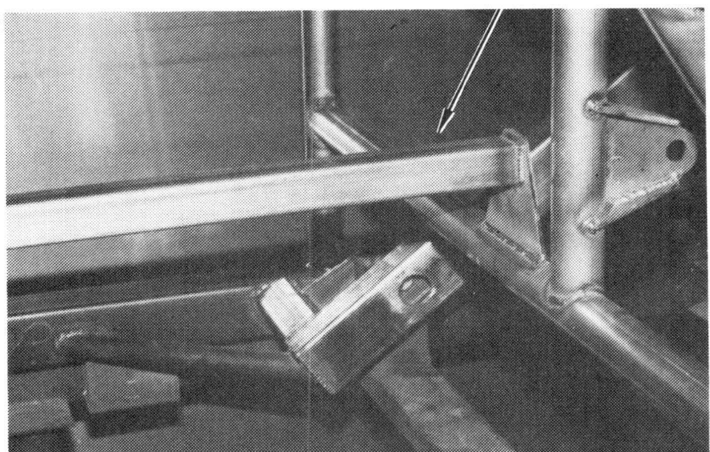

This track houses the slider bracket which adjusts the effective length of the torque arm. Below, another way of adjusting the effective length.

The traditional mounting of the Panhard bar with it mounted to the housing on the left and the chassis on the right.

Torque Arms On Asphalt

Torque arms can be effective on asphalt and overcome some of the drawbacks of the three point system, but they can offer some problems of their own if not carefully designed and implemented.

High cross weight, when letting off the throttle going in to a turn, makes a real difference on how the torque arm reacts on the chassis. If the front of the torque arm is located right on the left-to-right center of gravity, the effect of letting off the throttle and getting on the throttle will effect the chassis evenly. However, if the torque arm is offset to the right of the lateral CG, the right rear tire is loaded under acceleration and the left rear tire is loaded under deceleration.

The major effect of a torque arm is under entry to a turn. It is an abrupt change. On asphalt, the torque arm has to be offset to the right to load the left rear under corner entry to keep the car tight, but on acceleration this torque arm location loads the right rear which creates an oversteer on turn exit. A left mounted torque arm on asphalt would load the left rear under acceleration, but unload it under deceleration and turn entry.

The reactions of the chassis on dirt are the same, but the transient reactions can work to the positive for this type of track surface. You want the rear end to be loose and kick out on turn entry on dirt, so the torque arm mounted to the left of the CG can accomplish this. On acceleration off the turn, the left mounted torque arm will load the left rear tire which will create more bite off the turn.

Lateral Locating Linkages

The Panhard Bar

The full length Panhard bar is the simplest and most widely used of the lateral control linkage systems. It is attached to the chassis at one side of the car and attached to the axle housing at the opposite side of the car. It is simply a long tube with spherical rod end bearings at each end which braces one side of the car against the other to result in no side shift of the axle.

A drawback to this system, as compared to the Watt's linkage, is that it moves in an arc rather than straight up and down. The shorter the bar, the sharper the arc it scribes.

Using a full length bar which goes all the way from one side of the chassis to the other, the amount of arc it moves in is minimal if the wheel moves a normal distance into bump (up to 2.5 inches). With more bump travel, however, the increased arc of the bar will begin producing significant bump steer.

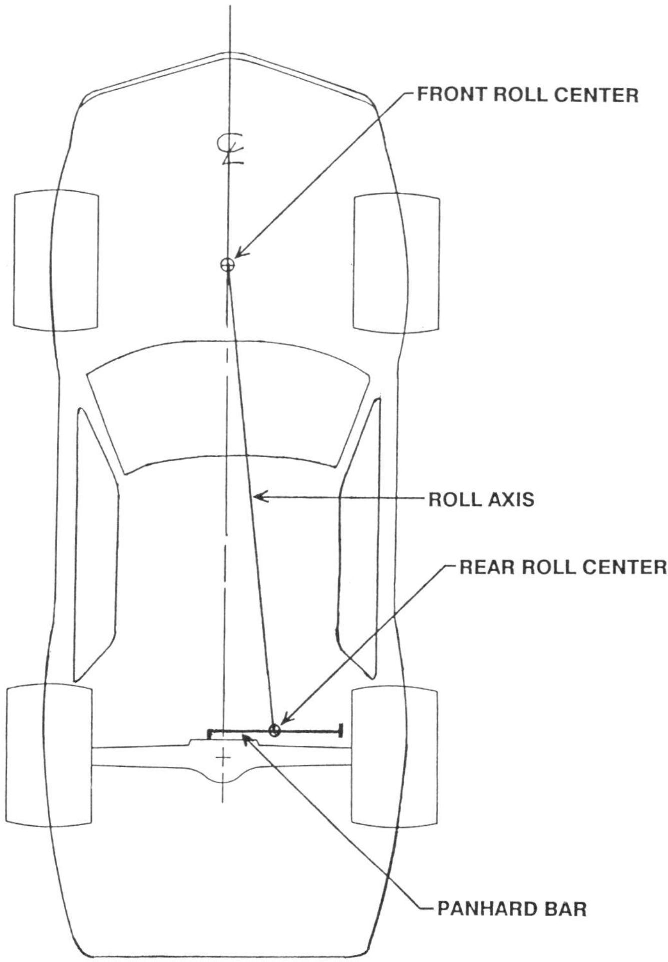

The short Panhard bar offsets the rear roll center and the roll axis.

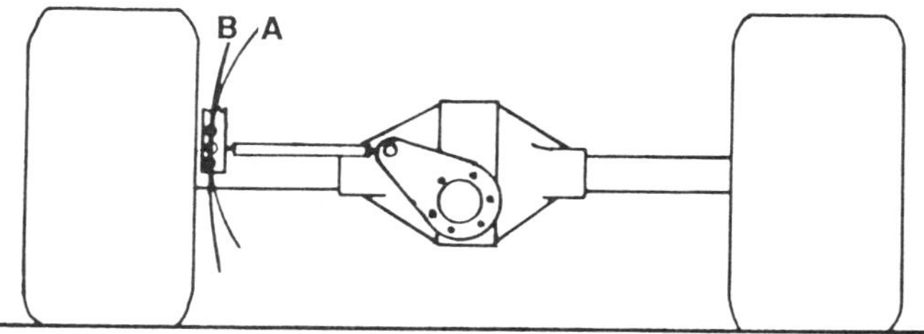

The short Panhard bar produces an arc (A) much sharper than the movement of the rear end housing (B).

The same problem is present with a short Panhard bar. Many applications mount the Panhard bar to a bracket off the housing center section around the pinion. The sharp arc of this short bar, as compared to the smoother arc of the wheel it is controlling, results in bump steer of the rear wheel. In one wheel bump on a bumpy track surface, this short type of bar also causes significant lateral tire scrub because of the difference in arcs. This results in increased tire heat.

The other problem associated with the short Panhard bar is that it moves the rear roll center from the mechanical center line of the car to a right offset position, which is a distance halfway in the middle of the length of the Panhard bar. This is undesirable.

Cars that use short Panhard bars usually have the bar attached to the right side frame rail, and use about a 15 to 20 percent stiffer left rear spring rate. Why? The right side mounted Panhard bar offsets the roll center roughly halfway between the vehicle center line and the right rear tire center. A stiffer left rear spring rate offsets the roll center back toward the left side, closer back to the vehicle center line.

The mounting of the Panhard bar is extremely important. First, it should be mounted parallel to the axle and to the ground. If the bar is mounted at an angle, its arc will move in a direction against the natural bump arc of the outside wheel, creating bump steer by forcing the whole axle toward the outside of the turn. Angling the bar uphill or downhill will cause the rear handling to be erratic. And you can see it with tire temperatures across the right rear tire. They will be very uneven.

It is extremely important that the Panhard bar be straddle mounted between two brackets which are stiff enough for the application. Most people do not realize the extreme torque which is placed on the mounting brackets of the Panhard bar. We have seen brackets completely ripped out of the frame rail because of improper mounting. The answer to this problem is to triangulate the mounting bracket with a cross tube.

The bracket itself must be of at least 0.25-inch thick material. If the bracket is not stiff enough, the outside wheel will experience deflection steer from the twisting of the bracket. Remember, the bracket may be strong enough, but still not stiff enough (meaning the bracket may be strong

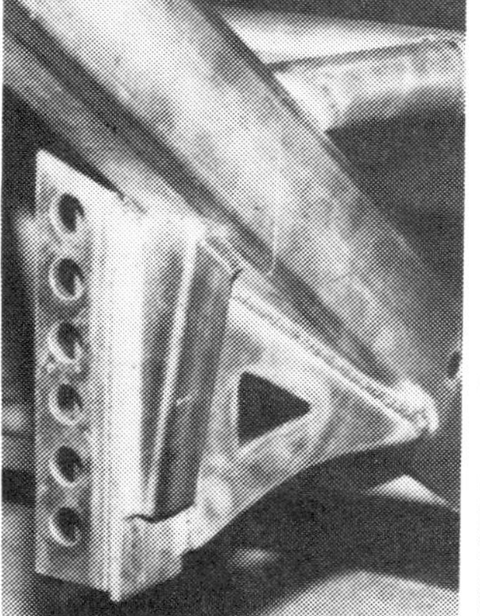

The bracket that mounts the Panhard bar to the chassis has to be stout. When there is no lower frame rail to help anchor it, it has to be braced in every direction.

The Watt's linkage for lateral control. Note how there is a lower frame cross member to help brace the frame brackets. Below, detail of the Watt's link center mounting.

enough not to break but still be weak enough to allow twisting deflections).

The Watt's Linkage

The ideal car has zero roll steer and bump steer which prevents it from feeling "twitchy" in turns and over bumps. A Watt's linkage is the only type of lateral control linkage which results in zero roll steer.

The Watt's linkage has a bellcrank which is attached to the center of the rear end housing on a pivot. It has two arms from the bellcrank, one extending in each direction, ultimately anchoring in a bracket on each side of the chassis. Each end of both these arms is free to pivot in spherical rod end bearings.

This system keeps the chassis centered in its proper position over the axle during all movements of bump and

Note the double shear brackets used to mount the Watt's link arm. These brackets are very strong, light, and easier to fabricate.

rebound, without any bump steer resulting from a linkage arm moving the outside wheel through a forced arc.

Rear Roll Steer

Roll steer is a change of directional angle of the rear end housing due to chassis roll. Roll steer is present if the rear axle moves from its perpendicular relationship with the vehicle centerline during chassis roll. Another way to picture it is that the wheelbase grows slightly longer on one side of the car, and slightly shorter on the other. It doesn't take a very big steering angle of the axle to make a significant change in handling.

On a solid axle (either front or rear), roll steer is caused by geometric differences in the arcs of each wheel (on the same axle) during body roll. One wheel is moved backward or forward relative to the other during body roll. This can happen because one side has a different length longitudinal control arm than the other side. Or, it also can occur if the left and right side control arms are symmetrical in length (which is most often the case) but the arcs of the two arms are different during bump travel compared to rebound travel. Remember that during body roll, one wheel (the outside one) is in bump while the other is in rebound.

The difference in trailing arm arcs is one of the most common ways that roll steer occurs. Remember that the housing-connected end of the trailing arm link moves in an arc as the car's body rolls. If the outside link starts with the back of the link at a downhill angle, that arc will lengthen the effective length of the link as it moves up in bump. Conversely, if the outside link starts with the back of the link at an uphill angle, that arc will shorten the effective length of the link as it moves up in bump.

To check for and adjust roll steer, the static angles of both the left rear and right rear trailing links must be addressed.

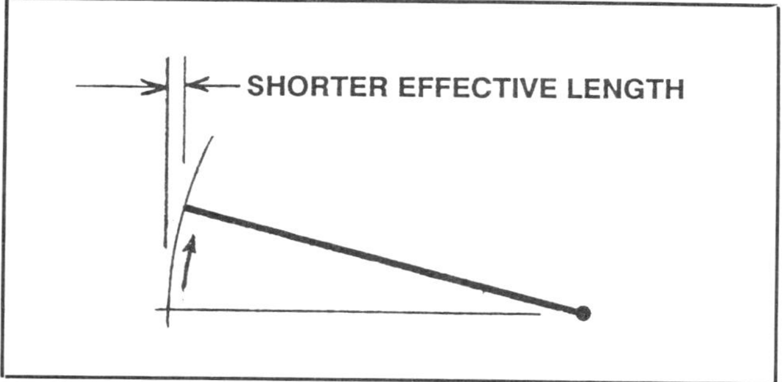

When the front of a link is mounted down hill in front, the effective length of the arm shortens as it moves up.

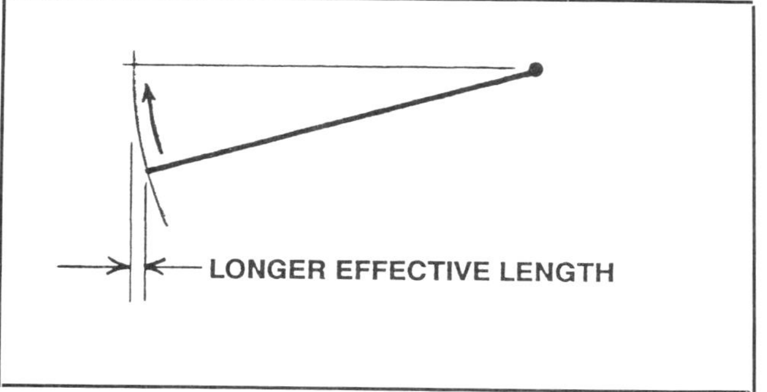

When the front of a link is mounted up hill in front, the effective length of the arm lengthens as it moves up.

The rear end can be kept square with the chassis during body roll if the right rear link length gain is matched by the same amount of left rear length gain. Or, roll oversteer can be achieved by link length gain at the right rear and a shortening of the link length at the left rear. Roll steer is strictly a factor of body roll. If there is no body roll, there is no roll steer. And, the more the car rolls, the more roll steer there is. It can be seen here that roll oversteer is at its greatest when the car is cornering the hardest (and needs it the most) at the apex of a turn.

It is best not to build in any rear roll steer. Make sure the car corners properly with neutral steer. Sometimes roll steer is present in a car without the racer knowing it. If you have a problem with oversteer or understeer from the middle of turn entry to just past the turn apex (the area of greatest body roll), especially on a fairly flat track, one of the items to check for is the existence of roll steer.

Using Roll Steer

First of all, measure your car through increments of bump and rebound travel on the inside and outside wheelbases, and determine the existence of roll steer in your car.

Roll understeer should be avoided. It is difficult to balance the chassis with it present. The only way to balance the chassis (without fixing the actual cause of the problem

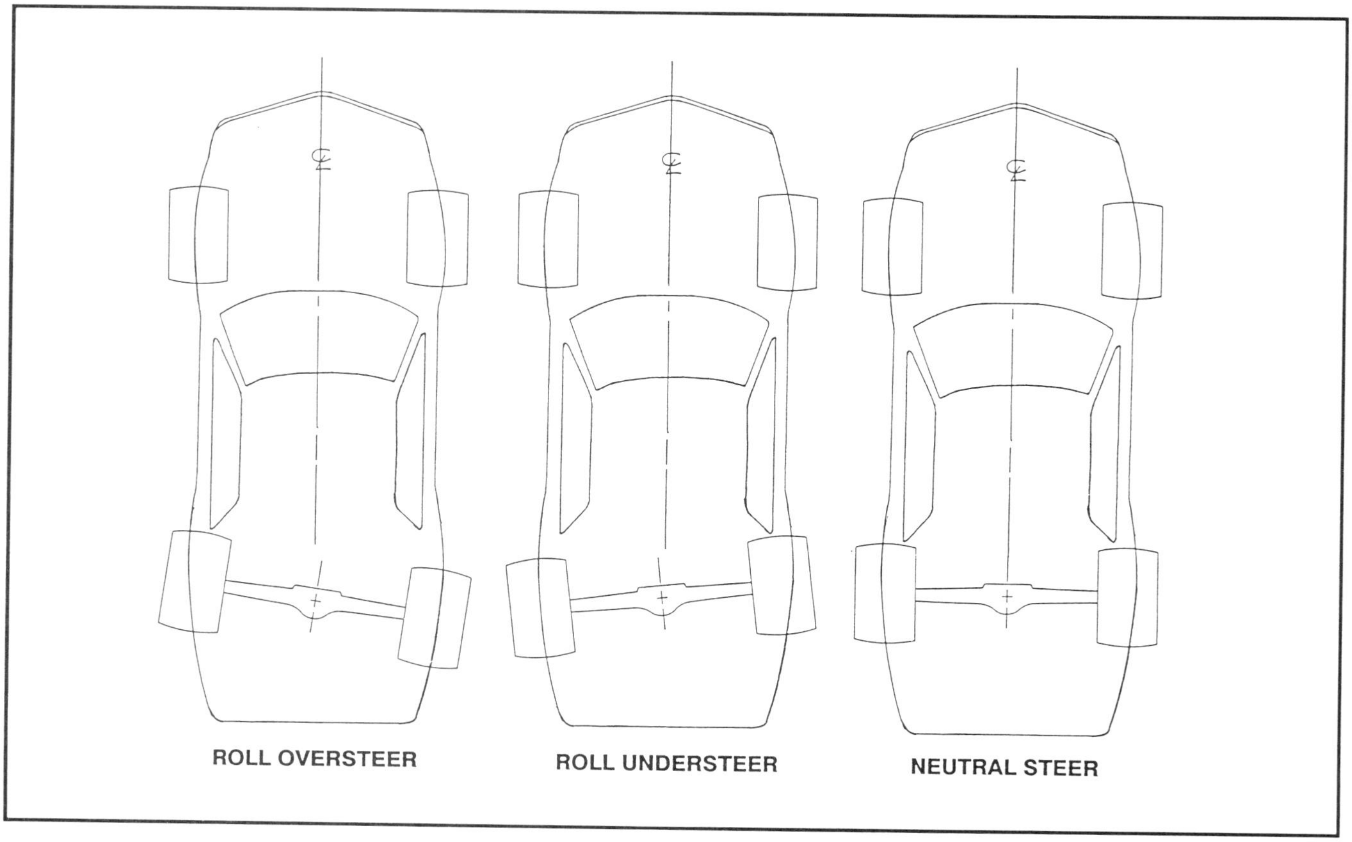

ROLL OVERSTEER

ROLL UNDERSTEER

NEUTRAL STEER

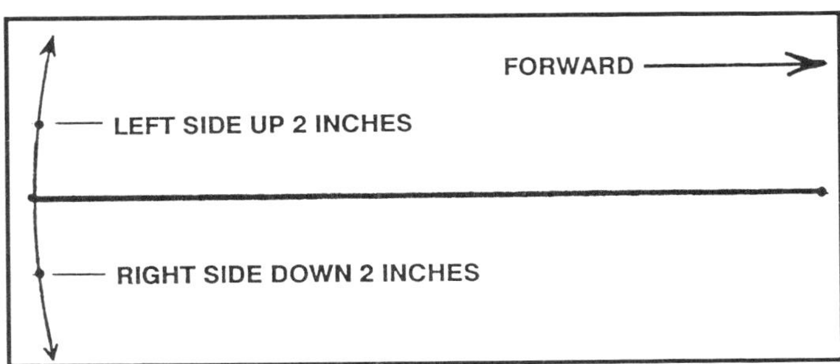

As long as the 2 lower links are equal in length and parallel to each other and the ground, no roll steer will be present because each side is pulled forward and equal amount during body roll.

with the suspension linkage) is to increase rear oversteer by adding too much rear tire stagger. But, this is introducing a new problem into the chassis to cover up an existing one.

A small degree of roll oversteer can be helpful, especially with a car which has a solidly locked rear end (spool). It works in a positive way the same as a small amount of tire stagger does. It can be very helpful on a dry, slick track, because the driver can change his corner entry method to control the rate of body roll. The rate of roll oversteer will be proportional to the rate of body roll. If he uses the diamond pattern of cornering, there will be only a small amount of body roll during initial turn entry and straight line braking. A higher front roll couple (more understeer) helps keep the car stable here. Then when he makes his turn to the left at the late apex, a sudden amount of body roll is induced and with it comes a sudden amount of roll oversteer, which helps him cleanly make the turn. Then its a straight acceleration off the turn, with body roll rapidly decreasing.

Some roll oversteer, in general, is good for dirt track cars for the reason that it lessens the slip angle of the right rear contact patch to the direction of travel as the rear end kicks out. If you oversteer the car with roll steer, you don't have to get the car sideways. You are getting the center of the tire contact patch back parallel with the direction of travel of the car with roll oversteer. It takes slip angle out of the tire contact patch which creates more forward bite and thus more traction. Good forward traction keeps the rear end tighter rather than on the ragged edge. The roll oversteer helps the tire never break traction.

The Floater Rear End

In the interest of safety, a full floating rear end should be used in any race car. Full floating means the axles transmit torque only, and they do not bear any of the vehicle's weight. If an axle should break in a full floater, the wheel will not come off as would happen in a semi floating rear end as found on passenger cars. The full floating rear end will cost more money to purchase than a non floater will, but it is an investment that will pay you dividends manyfold in terms of durability and safety.

The Spool and the Detroit Locker

The normal passenger car differential applies driving force to the left and right rear wheels in a turn at different rates. More force is delivered to the more heavily loaded outside wheel so that the inside tire is not scuffed. This has no relativity to a racing situation however. A race car needs to have both left and right tires hooked up solidly under acceleration. This is accomplised in race car with either a "spool" or a Detroit Locker.

There are several ways to lock a rear end. The worst way is by welding the spider gears solid. Welding on these heat treated high alloy steel gears can cause stress cracks in the metal, causing a failure under operating conditions.

The best technique of locking a rear end is to use a machined spool to replace the spider gear assembly. A machined solid spool reduces weight and rotating inertia, and is the cleanest way to solidly lock the rear end.

With a Detroit Locker-equipped rear end, any time that overspeed or differentiating must take place, the inside rear wheel is disengaged and it rotates freely. Under straightaway conditions, the Detroit Locker solidly locks both rear wheels. The big advantage of the Detroit Locker is that it eliminates inside rear tire scrub found with solidly-locked rear ends.

The only major service problem with a Detroit Locker is that the springs inside it lose their tension after a while. The symptoms of this problem are that the car veers to one side because one axle is not locking up completely. To check the problem, disasemble the Locker and check both the height and the rates of the two springs on a valve spring checker.

The type of differential used has a bearing on braking stability. The solidly locked rear end will definitely prevent one wheel from locking up at the rear. With the Detroit Locker under deceleration, the most lightly loaded wheel is unlocked and runs free. This offers an advantage over the spool, because with the left rear wheel unlocked going into a corner, the car will turn to the left much easier. The heavier the total weight of the car, the more important this fact becomes.

The disadvantage of understeer caused by the use of the spool can be adjusted out of the chassis by adding more rear tire stagger. As a guideline, a track and chassis set-up that would normally require 2 inches of stagger with a Detroit Locker would require 3 to 3.5 inches of stagger with a spool.

The spring rod trailing arm, mounted on the right rear, compresses slightly under acceleration to promote understeer coming off a turn by allowing the right rear tire to move forward.

Chapter 6

Shock Absorbers

Basic Shock Absorber Control Parameters

Shock absorber ratio, such as 50/50, or 90/10, is a ratio of the control of the shock absorber unit as it is divided between compression (the "down" or bump stroke) and rebound (the "up" or extending stroke). A 50/50 ratio means that the damping control of the shock is equal in both directions. And, this is the most widely used control ratio of shock absorber. A 90/10 shock would have 90 percent of its total damping control on compression, and only 10 percent on extension. That means this is a shock that is very easy to pull out, but very heavily controlled on the compression. The most common use of a 90/10 shock on a Pro Stock sportsman car is as an axle damper unit, such as the PRO Shocks' 7910 unit, which has 600 pounds damping force on compression and only 65 pounds on the extension stroke.

The pounds of damping force rating is another important parameter. Not all 50/50 shocks react to input forces the same. Specifications for diferent brands of shock absorbers will show that they are rated at 260 pounds or 315 pounds, for example. What this means is the number of pounds of force required to cycle the shock absorber at so many inches per second. Most manufacturers show the rating of their shock absorbers at 24 or 25 inches per second, which is the most critical cycling and control range for a shock absorber on a race car. This is the velocity of the piston in the second control range in a three-stage shock absorber. It is this number that is usually matched to the wheel rates of a race car.

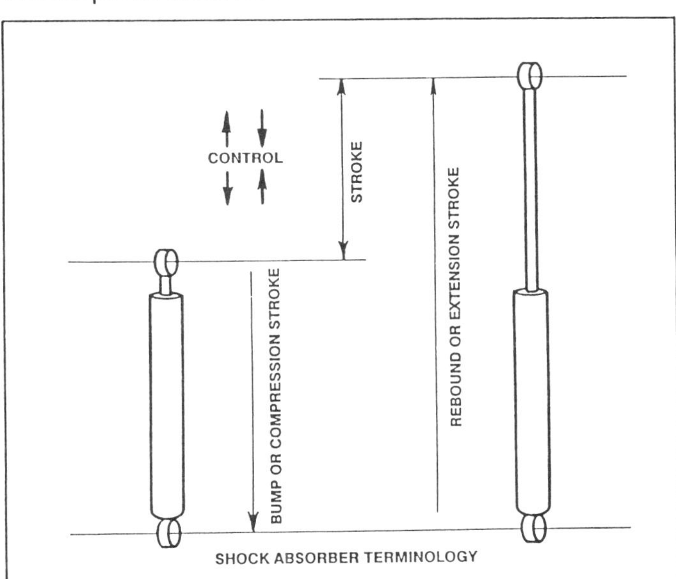

SHOCK ABSORBER TERMINOLOGY

Matching Shock Absorbers To Springs

Many successful pavement short track cars use shock absorbers that are rated softer than for the spring rate they are using. This is done so that the shock doesn't overpower the spring. This is contrary to thinking that has been prevalent for 30 years. It used to be the popular thinking that you had to have a shock stiff enough to control rebounding of the spring. But shock absorber valving design — in certain brands — has improved so much in recent years in the correlation between low speed control and high speed control that a specific shock can be used with a wide range of spring rates. For example, the PRO Shocks number 5 (second digit in the part number) valving is very effective on a paved track with wheel rates between 150 and 550 pounds per inch with no negative effects.

The current thinking with shock absorbers is to match the shock to the track and the body roll control needed, and not the spring. It also has to match the behavior requirements of the driver. A smooth driver can get away with less shock control because he doesn't fight the car, pitch the car around, or make radical moves. A smooth driver will allow a car to drift out of the line a few feet if he gets it in a turn too hard, instead of pinching the car down hard. This takes less shock control. Be aware of this — and be honest with driving style evaluation — when choosing shock rates. If your driving style causes more erratic loadings on the chassis, you'll need more shock rate.

Shock Absorber Part Number Cross Reference

(All Numbers Are For Steel Body Shocks)

AFCO	Carrera	PRO
1073	3173	7300
—	3172/3	—
—	3176/3	—
1074	3174	7400
—	3173/4	—
1074-6	3176/4	7460
1075	3175	7500
1075-3	3173/5	7530
1075-6	—	—
—	3177/5	—
1076	3176	7600
1076-2	3172/6	—
	—	7630
1076-4	3174/6	—
1077	3177	7700
1078	3178	7800
1079-1*	3171/9*	7910*
—	3174/9	—
1093	3193	9300
1093-5	3195/3	—
—	—	9380
1094	3194	9400
1095	3195	9500
1095-6	3196/5	9560
1096	3196	9600
1097	3197	9700
1098	—	9800

Notes:

1. On split valving shocks, AFCO and PRO specify the compression valving as the first number and the rebound valving as the second number. For example, a 1076-4 AFCO shock has a 6 valving on compression and a 4 valving on rebound. A 7630 PRO shock has a 6 valving on compression and a 3 valving on rebound. Carrera specifies their numbers just the opposite — rebound first and compression second.
2. To convert AFCO steel body shock part numbers to aluminum body shocks, instead of 10 at the beginning of the part number, add 11 for smooth body shock, or add 13 for threaded body shock
3. To convert Carrera steel body shock part numbers to aluminum body shocks, instead of 31 at the beginning of the part number, add 71 for smooth body shock, or add 61 for threaded body shock. To convert to steel body shocks with weld-on bearings (cannot be used for coil-overs), add 21 in place of 31.
4. To convert PRO steel body shock part numbers to aluminum body shocks, add A at the beginning of the part number and drop the last 0 for smooth body shocks, or add AC at the beginning of the part number and drop the last 0 for threaded body aluminum shocks.

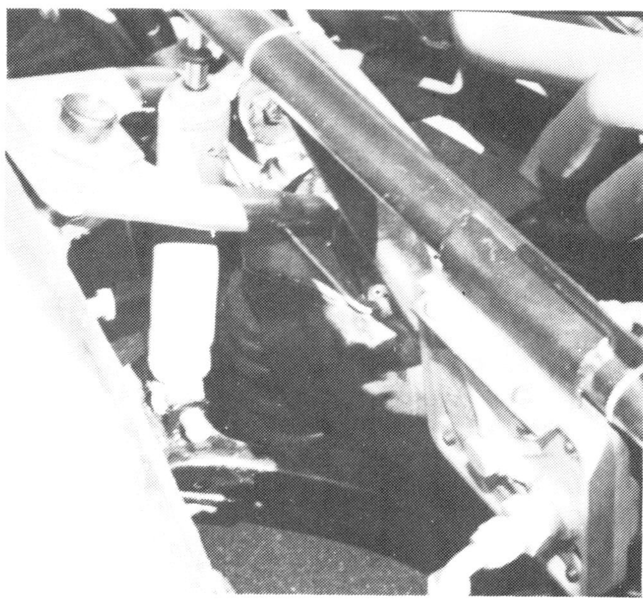

The proper mounting position for the front shock absorber is up through the middle of the upper A-arm. This gets the bottom mount out very close to the end of the lower A-arm, for more effective control. Below, a shock mounted to the side of the lower A-arm such as this gives very minimal shock control to the wheel.

Mounting Shocks

The more travel the piston in a shock has per increment of wheel travel, the more effective the damping will be, regardless of the rate of damping. This means that the closer the shock is mounted to the wheel, the more effective it will be. Make sure there is adequate clearance between the shocks and the tires through full suspension travel and steering movement. Shocks are most effective when operating nearly vertical. A static angle of 10 degrees from vertical is ideal, with the top of the shock toward the center of the car. The shocks on the front corners should be mounted so they move in line with the A-arm movement — not against it.

With the car at proper ride height, the shock absorber should be at one half of its total stroke travel. The stroke of the shock should be long enough so that the shock never reaches the end of its travel or bottoms out during the full extent of wheel travel. If it does bottom out on bump travel on the race track, the spring rate will immediately jump to infinity (solid suspension) and you will probably turn the shock into junk also.

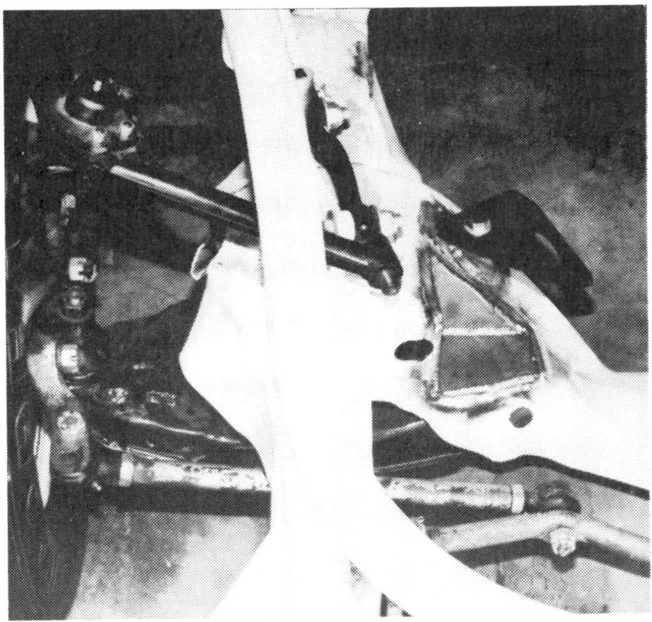

Mounting the front shock inside the A-arm requires the spring pocket to be trimmed for clearance.

The proper type of mounting end for a stock car racing shock is either a tie rod end or spherical bearing. The spherical bearing is easier to work with and provides a more positive mounting. Let your budget be your guide.

For the front end of a Pro Stock car, use 7-inch stroke shock absorbers, and use the 9-inch stroke shocks in the rear.

To plan the correct positioning of the shock mount on the chassis, set the frame and suspension at the desired static running height (fixed rigidly). Attach the lower shock mount to the suspension, bolt on the shock, and measure the total compressed length of the shock (say 14 inches). Compute one half of the shock stroke length (9 inches times .5 = 4.5 inches). Extend the shock to one half of its compressed length (14 plus 4.5 equals 18.5 inches). Then attach the upper shock mount at the point on the chassis where the shock falls at 18.5 inches total length from the bottom mount. This then gives you room for 4.5 inches of compression travel and 4.5 inches of extension travel for the shock. This is just perfect for the rear end of a stock car.

Rear shock absorbers should ideally be mounted 5 inches inboard of the brake rotor rear face, and 8 inches below the centerline of the axle housing tube.

The proper mounting position of the shock absorber in the rear is 8 inches below the center line of the rear end housing and 5 inches in from the rear face of the brake rotor.

Chapter 7

The Braking System

Why concentrate on a competitive braking system for a race car? Very simply, stopping a race car quicker in a shorter distance can be one of the biggest advantages a racer can have. Frankly, very few racers really concentrate on brakes as an area to improve speed. They would rather concentrate on bite off a corner, or horsepower for the straightaway. But a driver with a good braking system can overcome those other advantages on turn entry.

Braking Vs. The Tire Contact Patch

Utilizing the braking capability of a car is just as important as having a good braking system. A look at the traction circle theory (see the Tires chapter) shows that a tire has only a certain amount of traction capability in any direction. The maximum tractive force can be used up by accelerating, decelerating, cornering, or in a combination of any of the three. When hard braking is being experienced, the majority of the tire traction is used up by deceleration, leaving very little tire traction available for turning (see diagram). What this means is a driver shouldn't be hard on the brakes while trying to turn in to a corner. There just isn't enough tractive forces available to handle both situations. Braking should be done hard and quick in a relatively straight line before making a hard turn into a corner. The driver should be almost completely off the brakes when he makes the hard turn into a corner.

Braking System Hydraulic Basics

How the master cylinder feeds pressure through the braking system is a basic lesson in hydraulics.

The system begins with an actuating force fed into the master cylinder. This force is the driver's foot on the brake pedal. The most force a driver can generate is between 150 and 200 pounds. But most racing brake applications are about 75 to 100 pounds of force.

The driver is also limited to about six inches in the amount of pedal travel he can apply (depending on the driver's leg length, closeness to the pedal and leg angle). The ability of the driver to apply force to the pedal is determined by body restraints (seat belt and shoulder harness), and the angle of the hip, knee and ankle joints. The more these three joints are in a straight line, the easier it is to create high force on the pedal.

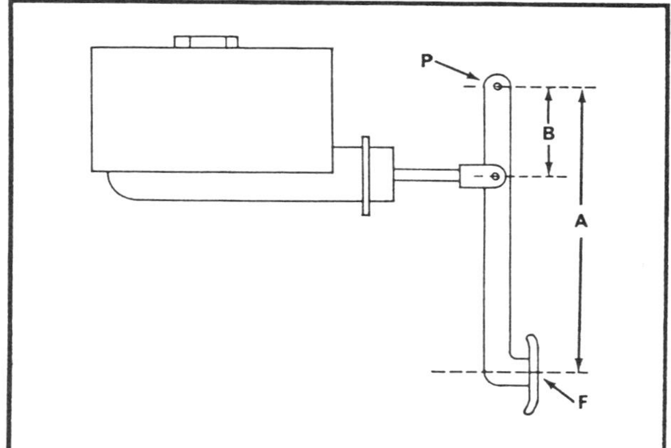

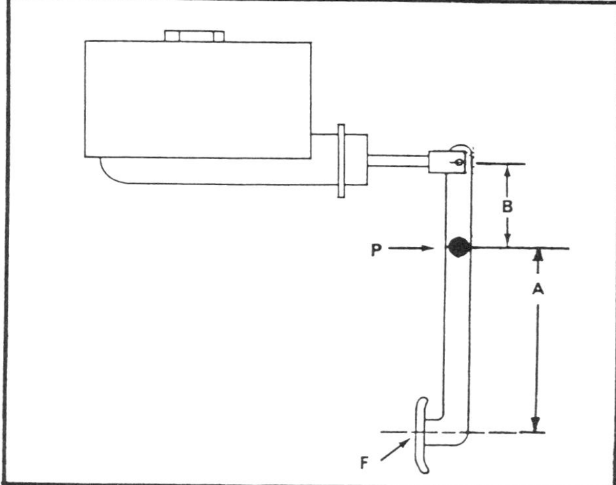

Pedal ratio is "A" divided by "B". "P" in the drawings is the pivot point of the lever arm and "F" is the input force. Pedal ratio is a method of gaining more brake pressure with mechanical advantage. As pedal ratio increases, pedal pressure is reduced, but pedal stroke increases.

The force from the foot is then multiplied by the leverage of the brake pedal lever on the master cylinder pushrod. On most passenger cars, this leverage ratio is between 3 to 1 and 4 to 1. A more suitable leverage ratio for a racing

Remote brake fluid reservoirs add extra capacity to the master cylinders to insure having enough fluid volume for the system.

vehicle is at least 5 1/2 to 1, with 6 to 1 being more common. This leverage ratio multiplies the driver's input force into the master cylinder. If 100 pounds of force is applied to the brake pedal and the leverage ratio is 6 to 1, the the input force into the master cylinder is 600 pounds (100 times 6). If the pedal ratio is increased, the input force will increase, but the pedal travel will also increase. A good idea would be to have alternate holes drilled in the pedal lever so alternate leverages can be applied — much cheaper than changing master cylinders. If you do this, remember to have alternate holes for both the pedal mount and the master cylinder pushrod so the pushrod is never applied at an angle.

The master cylinder bore size (and thus its square inch piston area) is the next consideration. The most common master cylinder size is one inch. This gives an area of 0.785 square inches (area is found by the formula A = .785 times D squared, where D is the piston diameter).

The brake system output pressure (measured in pounds per square inch or PSI) generated by the master cylinder is calculated by the formula: F divided by A, where F is force (the leg input in pounds) and A is area of the piston (calculated in above paragraph). For example, if a 1-inch diameter master cylinder is used, and the driver inputs 75 pounds with a 6 to 1 ratio pedal, the force is 450 pounds (6 times 75) and the area is .785, so 450 divided by .785 is 573.2 PSI system pressure. If a 7/8-inch master cylinder piston was used instead, the system pressure would be 748.7 PSI.

From this knowledge, it is easy to see that the smaller the master cylinder bore, the more pressure that is supplied into the calipers at the wheels. So why not use a smaller master cylinder piston diamter, such as 3/4 or 1/2-inch? There are two reasons against it: 1) It requires more pedal travel for a given fluid displacement, and 2) Higher line pressure aggravates line and caliper expansion and thus uses up more fluid. If a braking system is sufficiently stiff (meaning no or little line expansion or caliper deflection) that the driver is only using a small part of the total pedal travel available, then a master cylinder with a smaller bore would reduce the pedal force or effort required by the driver to stop the car.

Once the pressure leaves the master cylinder, many people believe that there is no great movement of fluid in the system. They believe there is only pressure transmittal. The pressure transmittal part is correct, but there is still fluid movement too.

Disc brake systems are set up with the pads riding about 0.002-inch away from the rotor, so many people assume that very little pedal travel is required to actuate the system. This is not so. Let's assume we have a four-wheel disc brake system which employs four pistons, each with a 2.75-inch diameter piston. Each caliper piston travels 0.006-inch (taking up clearance plus allowing for material compressibility). The volume of fluid required for each piston to travel 0.006-inch is 0.006 times the area of the piston (which is 0.785 times the diameter squared, or in this case, 5.937 square inches). 0.006 times 5.937 works out to 0.036 cubic inches. Multiply this by four caliper pistons, and we get a total fluid volume of 0.144 cubic inches.

How much piston travel (in inches) does this fluid displacement mean in the master cylinder? Assume we have a 1-inch bore cylinder, which is 0.785 square inch of area. Divide the fluid volume (.144) by the area of the master cylinder piston (.785) and we get 0.183-inch of travel of the master cylinder piston. If the pedal leverage ratio is 6 to 1, then this translates to a little more than one full inch of pedal travel (6 times 0.183 is 1.098). We must add to this pedal travel the pedal deflection and master cylinder deflection (assume 0.25-inch) and caliper deflection (assume 0.003-inch per wheel), which adds another 1.5 inches pedal travel. So, to just begin applying force to the rotors at each wheel, we have used up 2.5 inches of pedal travel! This, of course, is assuming very low amounts of deflection throughout the rest of the system. Can you imagine what kind of pedal travel would be required with a high deflection system?

Take note of how the brake lines are run here. This is a beautiful installation job on a short track car.

The Brake Lines

The brake lines function as the blood system of the braking system, and just as in the human body, there are several pertinent facts which should be remembered to permit good circulation.

For all fixed lines in the system, use automotive double wall (Bundyweld) tubing with 3/16-inch outside diameter. Any larger diameter tubing may expand and create a spongy pedal because of the high operating pressure of a racing disc brake system. Never use copper tubing— it is too soft.

You will notice that throughout this chapter the rigidity of the braking system is mentioned over and over. The brake lines play a big part here. This is important if you want the force you impart to the master cylinder to be delivered to the caliper pistons. If you use that force in expanding the brake lines or calipers instead of applying force to the brake pads, you cannot expect to stop the vehicle effectively, and the driver will merely use up all the available pedal travel. The pedal will bottom-out on the floor or the master cylinder piston will will bottom in its bore. By making the system as rigid as possible, you are going to get the work to the pads, where it belongs.

The layout of the solid tubing running from the master cylinder to each caliper should be carefully designed to prevent the presence of unnecessary bends, and loops which can trap air. And, the lines should run consistantly downhill from the master cylinder to the breaking junction of the flex lines to aid in line bleeding.

Use a brake line tubing bender (available at parts stores or tool stores) to make good, neat tubing bends. Bending by hand or over a mold will produce crimps in the tube.

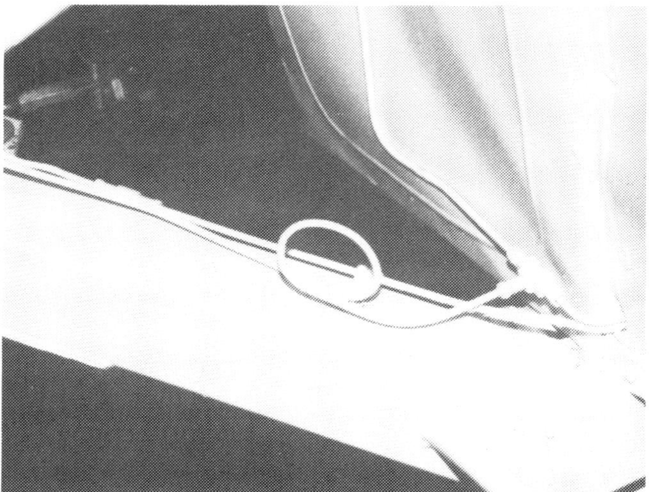

Free form lines and loops in the brake lines make it very difficult to completely bleed the system. It also makes the lines more susceptible to vibration damage.

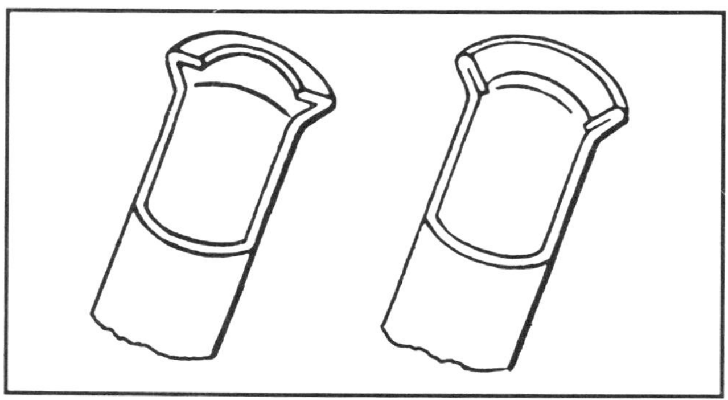

Steel brake line tubing should be double flared to prevent leaks under pressure. At left is the first operation of the double flare. At right is the finished flare.

The rubber-lined Adel clip protects brake lines, and electrical wiring too. The cushion assures tight fit and eliminates vibration.

Many people believe it is possible to proportion the braking force between the front and the rear brakes by installing larger diameter lines to the front wheels than the rear. This is NOT true. Pressure is independent of volume or area, so a 3/16-inch line to the front wheels will deliver the same pressure (PSI) as a 1/8-inch line to the rear wheels.

Never route brake lines near any source of heat, such as headers or exhaust pipes. Also, never route the brake lines in such a way as to have a loop or kink or high point where air can get trapped and defy bleeding techniques.

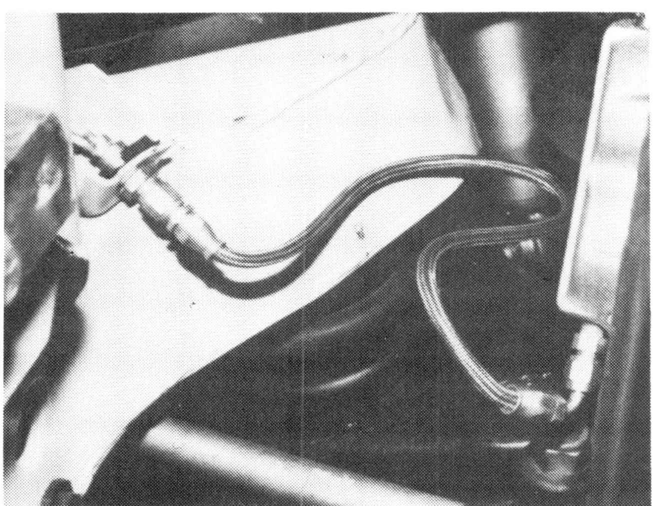

Make sure that the flex lines are of adequate length to accommodate full suspension travel plus full steering radius in both directions. Also note the solid frame mount that holds the connector fitting.

Steel brake lines should be secured to the chassis in several places along its run with rubber-lined Adel clamps. These aircraft-designed clamps prevent the transfer of vibration to the brake lines which could otherwise fracture the line material. All tee fitings along the steel tubing run should be secured to the chassis with a bracket.

At the connecting points of steel brake tubing, the fitting flares should be standard automotive double flares to prevent tubing fractures in the flares. And always cut the steel tubing with a tubing cutter, not a hack saw.

Flex Lines

At the point where the flex lines connect to the steel brake lines, the steel line must be secured to the chassis by a solid bracket mount. Otherwise, the input movements from the flex lines will also move the steel line, eventually causing a failure.

Only stainless steel braided teflon-lined flex hoses should be used to carry fluid from the steel lines to the calipers with the assurance that little pedal travel or work effort will be lost through line expansion. There are many brands of high quality plumbing lines available, but one of the finest on the market is made by Aeroquip. Use the "-3" (dash three) size hose, which has a 0.125-inch inside diameter, 1,500 PSI maximum operating pressure and 12,000 PSI burst pressure. When you buy flex line, be sure you know what you are getting — there are a lot of imitations out there that look like the real thing. There is also neoprene-lined steel braided hose available, but do not use it. Neoprene is not designed for use in high pressure applications such as the braking system.

To determine the amount of flex line to use at each wheel, jack the car up and let the wheel rest at full droop. Then install a line of adequate length which is not tight. For front wheels, steer the wheel hard to one direction as well as setting at full droop to determine the maximum length of line required. Always loop the front brake flex line excessive length. This acts as a spring to pull the line away from the wheel inner side.

A final thought about brake lines and flex lines: never assume that any brake lines or hoses you purchase are clean. Pour clean brake fluid through them just before final installation to be sure the smallest amounts of moisture, dirt and metal cuttings are removed.

Brake Bleeding

Once the braking system has been installed, all that is required is to add fluid into the master cylinders and depress the brake pedal to distribute it into the system. As the fluid travels through the system, it will displace the air which filled it. When the fluid reaches the calipers, there will be a large amount of accumulated air which must be released. This is the reason for the bleeder screw, which should be fitted at the highest point of the brake caliper.

To bleed the system, open the bleeder screw with a box wrench, attach a hose over the fitting, and run the hose into a jar containing NEW brake fluid. Be sure the tube is well covered in the jar with fluid so air will not be sucked back into the system.

Start with the farthest wheel from the master cylinder, which is usually the right rear, and progressively work to the closest caliper. Have an assistant press the brake pedal to the floor. Air and fluid will escape in spurts together. Make sure your assistant returns the brake pedal back very slowly. After the brake pedal has been depressed, close the bleed screw before returning the pedal. This prevents air

NEVER use rubber brakes lines in a race car. They don't have the wall strength to hold the pressures generated in a race car braking system.

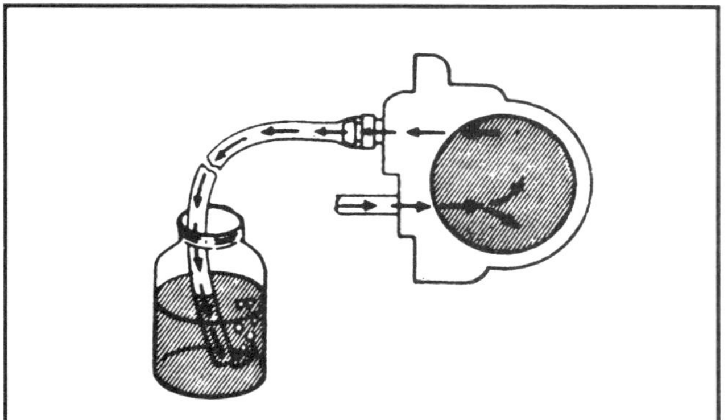

Typical bleeding operation of wheel cylinder into a jar.

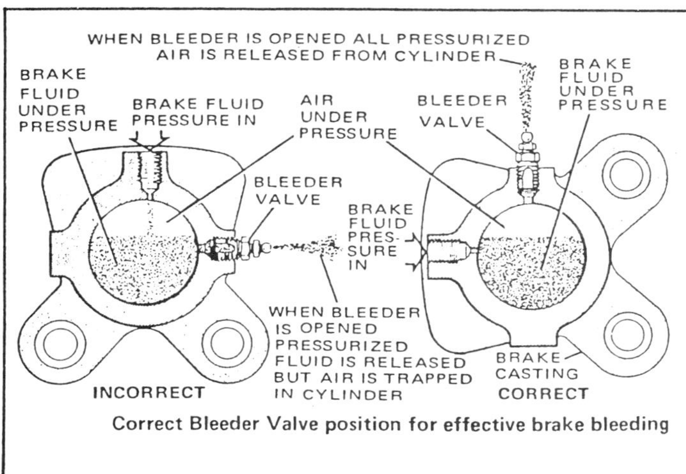

Correct Bleeder Valve position for effective brake bleeding

from being sucked back up into the system, and also tests the working quality of the master cylinder.

Never pump the brake pedal. This will aerate the fluid in the lines and you will never get the air out. Continue this process until all air has escaped and a solid column of fluid is observed. Then move to the next wheel. Be sure to check the fluid level in the master cylinder after each wheel has been bled.

With disc brake systems, air bubbles are difficult to dislodge sometimes after the initial installation of the calipers. Use a plastic or wooden mallet to tap the calipers gently to assist the air bubbles. It is also a good idea to bleed the brakes after a few practice runs have been made with a new or just-repaired system.

Before each racing event or weekly race, be sure to bleed the brakes, even if no air is dislodged, so the system can be replenished with fresh fluid.

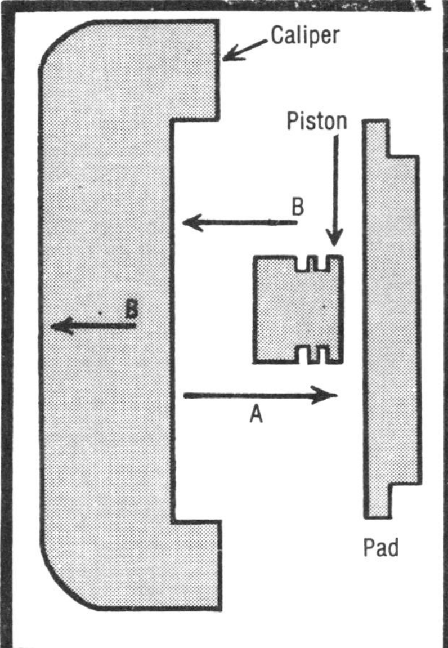

How caliper deflection can happen. Force from the hydraulic system is in direction A. The reaction to the force is in direction B which tries to deflect the caliper housing.

Brake Fluids

Brake fluid transmits force through pressure from the driver's foot to the brake pads. The basic premise which allows this to happen is that fluid is not compressible. Thus, it is able to transmit force.

The enemy of brake fluid, just like friction materials, is heat. If the fluid boils at any point in the system, or if the system leaks at any point, the fluid incompressibility is lost.

The important thing to keep in mind when purchasing brake fluids is that they are not all the same. The main ingredient is ethylene glycol, which has a lubricating capability for the system's rubber parts and is not highly susceptible to boiling.

Brake fluids in the United States are rated by the Department of Transportation (DOT). The DOT standards set minimum specifications and wet and dry boiling points for brake fluid. ALWAYS use a brake fluid that at least conforms to the DOT 3 standard, and much preferably exceeds it. Two brake fluids which do and are widely used by race cars are the AP 550 fluid and the Cartel 570 fluid. The numbers after the brand name refer to the minimum dry boiling point of the fluid.

Glycol-based brake fluids are highly moisture absorbant when exposed to the atmosphere. Because of this, it is important to buy brake fluid in small containers, such as a pint can, so the contents can be used up all at one time. The fluid in both the can and the master cylinder should be exposed to the atmostphere for the shortest possible time. The container in which the fluid is stored should be kept tightly sealed until the fluid is used. It is also a good idea to keep the master cylinder full, and be sure its cover has an expandable rubber boot which will take up air space as the fluid level decreases.

You should be aware that there is a DOT 5 standard for brake fluids. This is strictly for silicone brake fluids. These fluids have a higher boiling point, and as such have been promoted as a great racing brake fluid. Be aware, though, that it is not recommended for use in a race car. It contains several inherent problems: 1) its higher viscosity causes piston drag in the calipers, and 2) it is highly expansive under higher temperatures, causing more compressibility. Don't use it.

Calipers

The major item to consider in selecting components is caliper housing deflection. To explain why deflection is so detrimental to to the caliper, refer to the accompanying drawing. Remember that basic law of physics, "For every action, there is an opposite and equal reaction?" That certainly applies in the caliper housing. In the drawing, "A" refers to the initial action of the piston being forced against the pad, which in turn is forced against the disc. "B" represents the opposite and equal reaction. If the caliper housing was flexible enough, the system could use the rotor as the anchoring or back up point, and the fluid pressure would then apply force against the housing. This would deflect the caliper using all available pedal travel and deliver little if any stopping ability.

There are three basic items in the design of a caliper which determine how much it is going to flex: the material from which it is made, how thick the material is, and the bridge distance (or distance between clamping bolts).

Calipers are generally made from three common metals: steel, aluminum and cast iron. (There are also magnesium calipers, but that is out of the scope of the brakes used for this class of car.) The common denominator of all these metals, in relationship to their flex, is their modulus of elasticity. That is a number which relates the metals in their resistance to being deflected. The higher the modulus of elasticity number, the greater the resistance to flex. Aluminum has a modulus of 10 million, cast iron has a modulus of 14.5 million, and steel has a modulus of 30 million. In the aluminum variety, most aftermarket calipers are sand cast from 356 T6 aluminum. A pressure casting of alloy aluminum will provide greater strength per unit of weight.

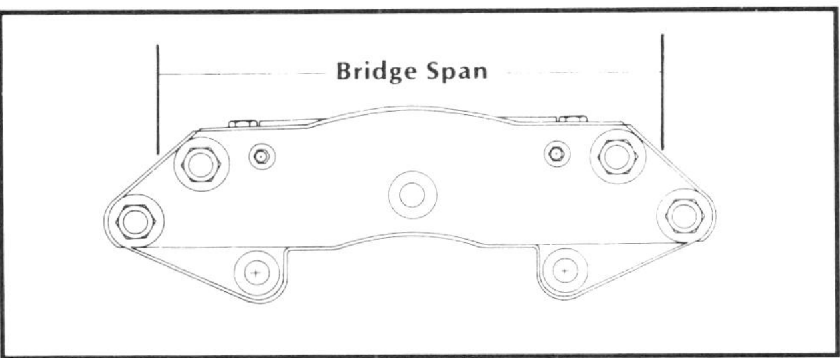

The bridge distance is the span between clamping bolts. The longer this span, the thicker the caliper housing material must be to resist deflection. Or, the longer this distance, the greater the deflection is going to be. So, the piston or pistons should be placed as close to the caliper clamping bolts as possible.

One of the biggest causes of deflection in a caliper is the use of inadequately sized bridge bolts. This is especially true in some aftermarket calipers of the two piece design. The largest possible bolts should always be used in the area of the bridge to minimize deflection. Only aircraft quality AN bolts and nuts should be used. It is important that the holes be reamed for proper fit and that the proper grip length bolts be utilized to eliminate wear and sloppy fits.

The question of bridge span is closely related to the piston size and arrangement. The simplest arrangement is one piston on one side of the rotor. Then there is one piston on each side, two pistons on each side and three pistons on each side of the rotor.

The best arrangement of pistons is two in a row on each side of the rotor. This spreads the force loading closer to the clamping bolts, putting it in combined tension and bending against the bolts, reducing caliper deflection. Three pistons in a row is not a preferable design because it places the center piston in the key stress area of the caliper housing. It puts a hole in the caliper body where solid material ought to be to prevent housing deflection.

Floating Vs.Fixed Calipers

The fixed caliper is one which has pistons on each side of the rotor, squeezing the rotor from each side. Because hydraulic pressure is always self equalizing throughout the entire system, one piston cannot overcome any others. Nothing moves in the caliper except the pistons pushing

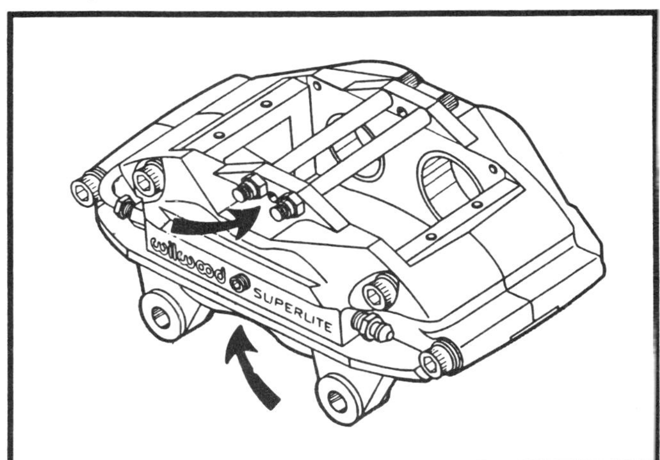

A fixed caliper has pistons on each side of the rotor, squeezing the rotor from each side. Thicker material in the housing between the bridge bolts and external stiffening ribs, as well as double center bridge bolts, add tremendously to caliper body strength and rigidity to resist deflection. Image courtesy of Wilwood Engineering.

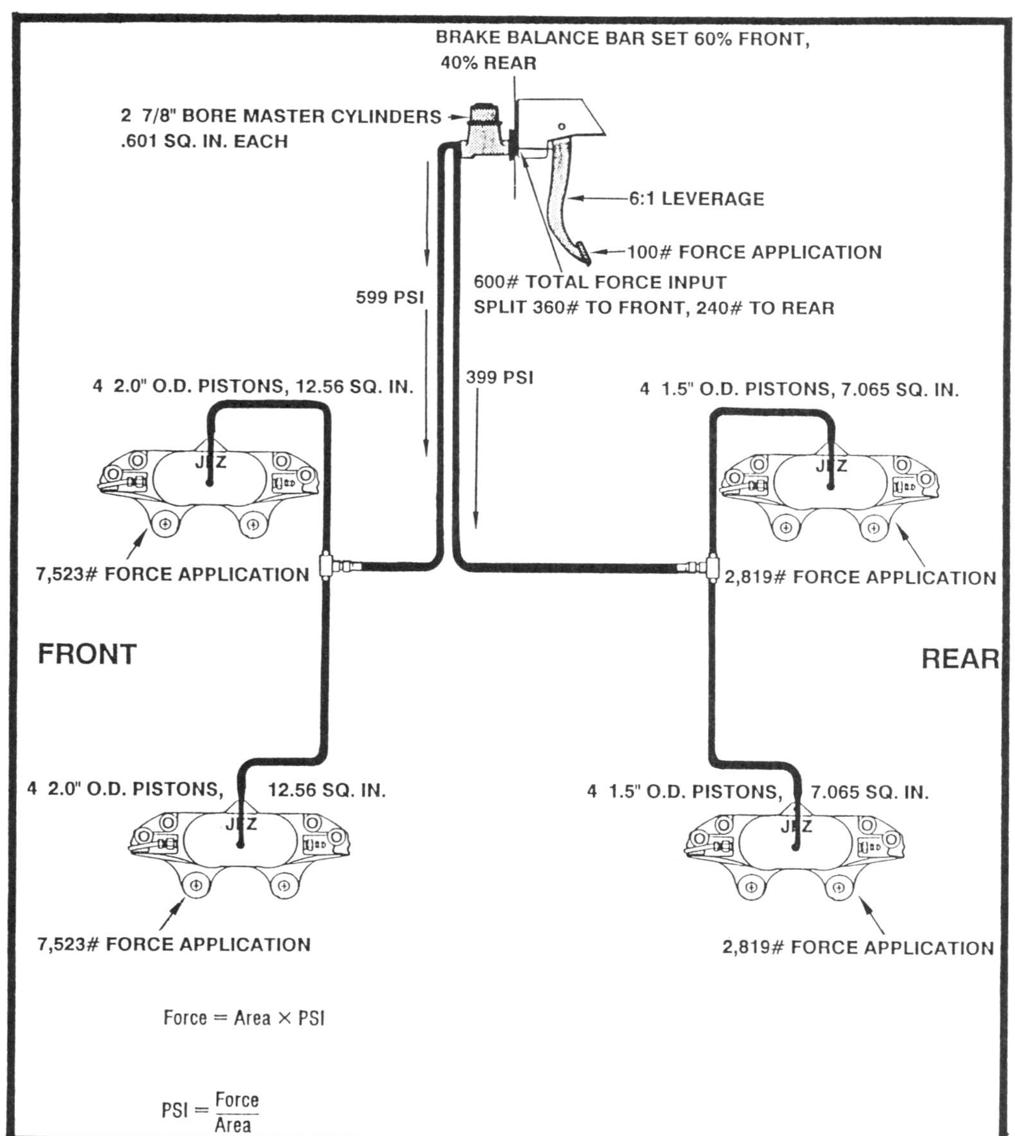

the pads inward from each side. It should be apparent here, then, that the critical strength item is the caliper housing. The housing must be absolutely rigid so the pistons can deliver their force to the pads and not deflect the housing instead.

The floating caliper was designed by passenger car manufacturers essentially to make the caliper less expensive to produce. It successfully applies the physics principle of "every action causes an equal and opposite reaction." Applying this principle, they eliminate pistons on one side of the rotor. The floating caliper is not solidly mounted but rather slides back and forth slightly on bushings. The piston travel on the primary side of the rotor is reacted by pulling on the housing so the pad on the secondary side is pulled tight against the rotor.

The floating caliper housing must be very rigid and of low deflection, or the entire principle behind it is lost, and so is braking ability. This is why the most effective full floating calipers are made of steel or cast iron — it offers a high modulus of elasticity.

If you use floating calipers, be sure to periodically check the condition of the bushings which the caliper floats on. If the sliding movement binds up on the bolt, there will not be full force application against the rotor.

There are advantages and disadvantages for both the fixed and floating types of calipers.

Fixed calipers are generally the type preferred for all-out racing conditions. The advantages are more pistons and piston area, and less flex. Floating calipers are generally smaller and lighter. The disadvantages: floating calipers tend to flex more, and their pad area is wider than and overlaps the piston face. This bends the pad and gives an uneven pad loading against the rotor. The multiple pistons — usually found in fixed type calipers — minimize this problem and give better pad clamping application against the rotor.

One of the biggest problems with floating calipers is not their tremendous frictional loss in contacting the outside pad, but rather the bridge design. On floating calipers, the bridge covers the top of the housing only — there is no material wrapping around the ends. This adds greatly to deflection and causes caliper clamping (after it heats up) even in the brake-released condition, robbing cool down time for the entire system.

Which Calipers And Master Cylinders To Use – Paved Tracks

Asphalt tracks demand the ultimate in race car stopping ability. Only the very best will do. If you had your choice, you would choose a fixed type caliper, light in weight made of high quality aluminum. However, many times track rules or budget constraints limit what can be used. There are several good companies making fine products — JFZ, Wilwood, etc. Get all the catalogs, read, them, talk to the manufacturers, and make your own choice. All make a wide selection of sizes. To help you choose the correct system for a typical Pro Stock 2,900 to 3,200 pound car, we have chosen to illustrate as an example the JFZ XL series calipers.

Braking Force Calculation – Paved Tracks

Our example system for a paved track car will use the JFZ XL200 calipers (four 2.0-inch O.D. pistons per caliper) on the front wheels and the JFZ XL150 calipers (four 1.5-inch O.D. pistons per caliper) on the rear wheels. Use two 7/8-inch bore master cylinders, one for the front calipers and one for the rears. Connect the pedal to the master cylinders on a balance bar with the proportioning initially set 60 percent to the front and 40 percent to the rear. Use a 6 to 1 pedal ratio, and we will assume a 100-pound input into the pedal from the driver's foot.

First figure the force input into the master cylinder. 100 pounds times 6 (the pedal ratio) equals 600 pounds. Multiply that by 60 percent to get an input of 360 pounds to the front master cylinder, and by 40 percent to get an input of 240 pounds to the rear master cylinder.

Next figure the master cylinder area, using A = .785 times D squared (A is area and D is piston diameter). In this case, A equals .785 times .875 squared, or .601 square inches for both master cylinders.

The system pressure is figured next, using Pressure = Force divided by Area. The force is 360 pounds for the front and 240 pounds at the rear, so 360 divided by .601 equals 599 PSI (front) and 240 divided by .601 is 399 PSI (rear).

The caliper piston area is the next item to be computed, using Area = .785 times D squared times the number of caliper pistons.

For the front, .785 times 2 squared times 4 equals 12.56 square inches. At the rear, .785 times 1.5 squared times 4 pistons equals 7.065 square inches.

Finally the output force (which is the force exerted by the pistons to the pads) is found by multiplying the system pressure by the caliper piston area. For the front, 599 PSI times 12.56 square inches is 7,523 pounds of force. At the rear, 399 PSI times 7.065 square inches is 2,819 pounds of force.

These output force numbers shown above are very typical for a paved track Pro Stock car. If you are choosing components to design and engineer your own braking system, use the steps we have shown above to calculate the output force you will have at the caliper pistons. Your numbers should be close to the ones shown here.

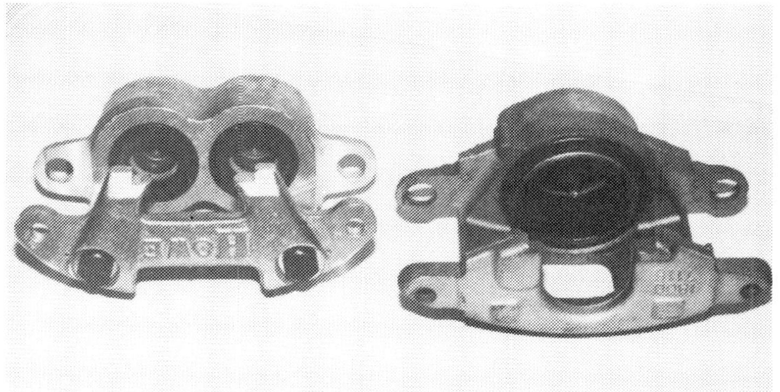

Two different Howe floating type calipers. At left, 2 pistons are enclosed in the same area as the GM sized caliper as at right, giving a much larger piston square inch area (7.09 versus 6.77 for the single piston).

Note: Many Pro Stock racing association rules indicate that only OEM type of brake calipers can be used, or if aftermarket components are used, a 100-pound weight penalty is assessed to the car. If you face that situation, add the better aftermarket brakes to your racing budget and take the weight penalty. Your overall performance on the track will be much improved with superior brakes.

Which Calipers And Master Cylinders To Use – Dirt Tracks

A race car does not need to generate a lot of braking torque on dirt tracks. In fact, too much braking force is detrimental. If you use the type of braking system just illustrated above for an asphalt track on a dirt track car, you will experience severe understeer on corner entry and probably lock up all four wheels.

A different type and size of calipers is required on dirt. A race car on a dirt surface requires more rear braking force. With more rear brake, if a car gets into a pushing situation, the driver can just stab the brakes and break the rear end loose. He can use the brakes and gas to steer the car and keep the rear end loose, keeping the car in the low groove.

A good choice for dirt track applications is the GM-style of floating caliper, either in steel or aluminum construction. A typical choice to illustrate this type of brake would be the Howe 337 for the front wheels, which is an aluminum caliper with one 2.9375-inch diameter piston, and the Howe 336A for the rear wheels (aluminum, with one 2.375-inch diameter piston).

Braking Force Calculation – Dirt Tracks

With the two specified Howe calipers, we are using a 1-inch diameter front master cylinder, a 7/8-inch rear master cylinder, a 6 to 1 ratio pedal mounted on a balance bar with a 60/40 front to rear bias distribution.

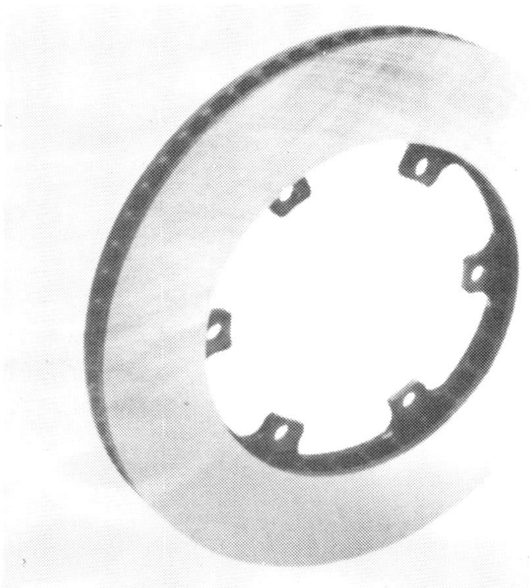

For the weight of a Pro Stock car, the minimum rotor size should be 1.25 inches thick and 11.75 inches in diameter.

A low bucks approach to consider is passenger car rotors from a wrecking yard. This one is from a 1969 Lincoln Mark IV. It is 1.190 inches wide and has a diameter of 11.72 inches. The bolt pattern is 5 x 4.5, but it can be redrilled for a 5 x 5 bolt pattern. Another rotor to consider is from a 1971-1973 Chrysler Imperial, which is 1.25 inches thick and has a 11.75-inch diameter. It has a 5 x 5 bolt pattern.

Force input: 100 pounds driver input times 6 (ratio) equals 600 pounds, split 60/40 means 360 pounds into the front and 240 pounds into the rear master cylinders.

Master cylinder area: front equals .785 times 1 squared equals .785 square inches. Rear equals .785 times .875 squared equals .601.

System pressure: front equals 360 divided by .785 equals 459 PSI. Rear equals 240 divided by .601 equals 399 PSI.

Caliper piston area: front equals .785 times 2.9375 squared times 1 equals 6.77 square inches. Rear equals .785 times 2.375 squared times 1 equals 4.43 square inches.

Output force to pads: front equals 459 PSI times 6.77 square inches equals 3,107 pounds. Rear equals 300 PSI times 4.43 square inches equals 1,768 pounds.

These numbers represent a typical dirt track Pro Stock braking system. When choosing components and engineering your system, use these same methods to calculate your system. The numbers should be very close.

Note: If your dirt track surface dries out very early and you end up racing the main event on a dry/slick surface, you might consider using aftermarket calipers that give you better stopping power. A good example would be to use the JFZ XL175 calipers (four 1.75-inch O.D. pistons per caliper) on all four wheels. Start the brake proportioning bias with at least 60 percent to the rear on a wet track with this set up.

Some Tips For Working With GM-Style Calipers

The GM-style of floating caliper, no matter who manufactures it, has an expansion problem. The aluminum calipers are much worse than the steel calipers. After they get hot and are deflected for a while, they will expand and take up all the clearance between the piston and pad and rotor. This will cause a pad drag on the rotor which in turn causes heat build-up. A tip to help remedy this is to gain more clearance in the system by milling or grinding approximately 0.080-inch off the piston face. And, never consider using an aluminum GM-style caliper on a paved track car.

Rotors

The primary design parameter for any disc brake system is the ability of the system to dissipate heat. It is the limiting factor of any disc system — and the ability of the caliper to generate a clamping force. The temperature of the rotors and thereby the entire braking system is critical to maximum braking performance. If the rotors cannot dissipate heat quick enough, the brake fluid can boil and total system failure can result.

Rotor thickness and diameter are both critical for obtaining proper heat dissipation and increasing overal braking effectiveness. The larger the rotor diameter, the greater the leverage exerted by the caliper on the rotating mass (axle

When installing the caliper brackets, the rotors must be directly in the center of the caliper or else the brake pads will drag. To get an accurate measurement, the brake hat must be bolted solidly to the hub.

or spindle). This translates into more stopping force. And, because larger diameter rotors have greater surface area, they will dissipate heat better. Likewise, increased rotor thickness will absorb more heat and help dissipate it quicker. The rotor is what dissipates the heat and energy created by braking. If the rotor temperature exceeds the thermal barrier of the brake pads, you loose braking ability.

For the Pro Stock type of braking system, a 1.25-inch thick rotor should be used, and the minimum diameter should be 11.75 inches.

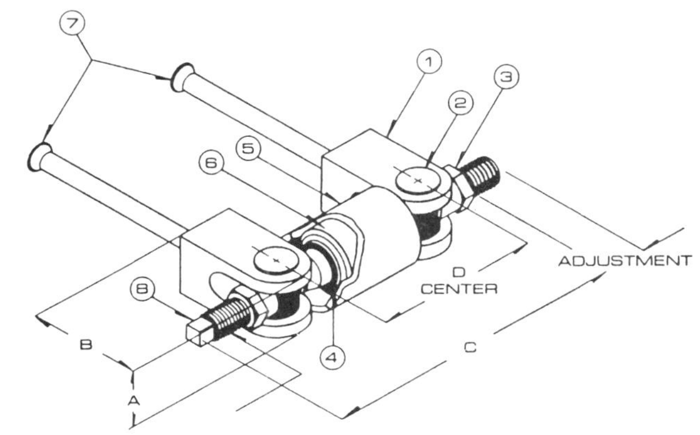

This is the balance bar assembly available from Tilton Engineering. A blueprint for building it is also available from Steve Smith Autosports (item #DB6).

Brake Balance Bars

A brake balance bar is a part which operates two master cylinders with one pedal. The pedal can apply force against an adjustable bar which inputs more force into one master cylinder than the other.

A brake balance bar can change brake proportioning to favor the front or the rear brakes. If a car pushes during braking/corner entry, add more rear brake bias. This restores more cornering traction to the front tires (refer to traction curcle theory in the Tires chapter). Be sure to make brake bias changes in small increments — you don't want to make a big change and then have the car swap ends when you hit the brakes at the next corner.

The brake balance bar should be used as a fine tuning tool only. It should not be used to compensate for the use of the wrong size master cylinders. The balance bar is a great tool for fine tuning the chassis, and to compensate for changing track conditions. But it is not intended to compensate for large inequities in the braking system.

To calculate master cylinder input force with a balance bar, the total pedal effort must be divided by the proportion of the effort being fed into each master cylinder pushrod. For example, a pedal has a ratio of 6 to 1, and the driver exerts 100 pounds of force on it. That's a total of 600 pounds input force. If the balance bar bias is set 50/50 to the front and rear master cylinders, the force is 300 pounds (50 percent times 600 pounds) to the front, and the same to the rear master cylinder. If the balance bias is 60 percent front, 40 percent rear, the force at the front master cylinder pushrod is 360 pounds (600 times 60 percent) and 240 pounds to the rear (600 times 40 percent).

Always calculate your balance bar initial set up with a 60 percent front, 40 percent rear setting, for both asphalt and dirt tracks. For paved tracks, that should get you close to the required adjustment. For dirt tracks, move the adjustment to favor rear braking bias (about 55 to 60 percent rear) to start for a wet dirt surface, or you'll find your car nose first into the wall the first time you brake into a turn. As the track dries out, continue to move the adjuster to bias the front brakes. For a dry/slick dirt track, a 60 front/40 rear bias is about ideal.

Balance Bar Adjustments

Master cylinder push rods should be adjusted so that when the pedal is retracted, both push rods are pulling against the retaining washers in the master cylinder. Any advancement of the push rod when the pedal is fully retracted will seal off the return bleed line and will lock the brakes on when the system heats up.

Master cylinder push rods should be adjusted so that when the brakes are on, the balance bar makes a right angle to the pedal. The balance bar should be parallel to

The easy way to locate the holes in the pedal mounting frame. The holes in the mounting bars should be sleeved to retain their rigidity. Instead of cutting and welding the bars, have them bent. This will prevent injuries to the driver's legs in case of an accident. Make sure the bars are welded at both ends.

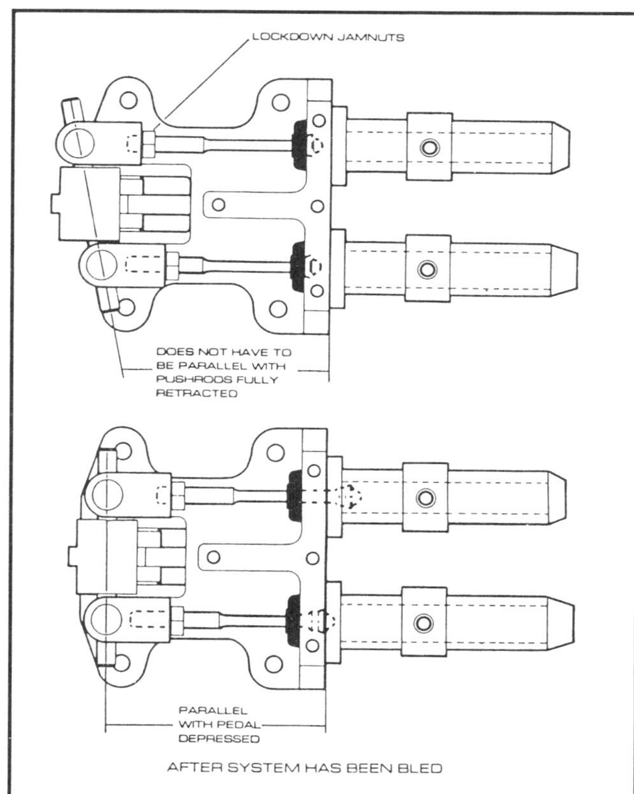

The clevises of the balance bar do not have to line up with each other when the push rods are retracted. However, they should be parallel when the pedal is depressed.

the firewall of the car in the brakes-applied position, NOT in the fully retracted position.

A proper balance bar has a floating spherical bearing inside of a tube for its pivot and uses clevises with a barrel nut attachment to the push rod. Balance bars that use a spherical bearing fixed to the pedal and rod end bearing attached to the push rod are not proper as the balance bar will not stay in a horizontal plane when apllied.

The length of the push rods in relation to each other have no bearing on the proportioning of the balance bar. The only effect the balance bar has on the leverage applied to the master cylinder is the relationship between the spherical bearing and the clevises. The clevis that is closest to the spherical bearing has the greatest force being applied to the master cylinder.

Other Brake System Tips

Always match brake pad compounds on the front and rear brakes. Mismatching pad compounds front to rear can cause problems when the brakes are coming up to operating temperature, as their coefficient of friction will be different earlier than later. This can cause a very unstable braking situation.

Be sure that brackets which attach the calipers are installed absolutely parallel to the rotor running direction. Brackets which are out-of-parallel will cause a loss of braking torque because effort will be used by the caliper trying to square itself.

If you use a pedal arrangement that mounts the master cylinders on the floor, be sure to use a 2 PSI residual check valve in the master cylinders to prevent all of the fluid in the system from draining back into the master cylinders. However, if you have top-hung or firewall mounted master cylinders, do not use a residual check valve in the master cylinders.

Master cylinders and pedals must be securely mounted on a body or chassis structure where pedal force cannot move the master cylinder mounting. If the mounting is not rigid, it has the same effect as having the brake lines expand.

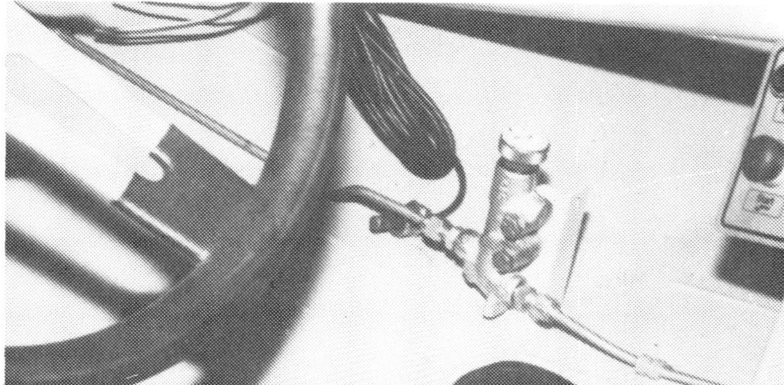

The Kelsey Hayes manually adjustable proportioning valve is installed in the line running to the rear brakes. It is located close to the driver for onboard adjustment.

Chapter 8

Tires

The vital link in any chassis set-up is tires. Tires, and more accurately, the tire contact patch or footprint, are the only connection a car has with the track. Racing tires have done more to improve lap times, as technology develops, than any other single factor. Racing tires have reached a high state of development and are technically complicated. However, a basic understanding of what goes on is necessary to make sound choices about tire compounds, sizes, use, pressures, and how they affect handling. Let's look at some of the basic characteristics of the racing tire.

Coefficient of Friction

Any time two surfaces move in contact with each other, a coefficient of friction exists. The higher the coefficient of friction, the more friction there is present. In a case of a bearing within an engine, a low coefficient of friction is desirable. In the case of a racing tire operating on a track surface, the highest possible coefficient of friction is desirable. The higher the coefficient of friction, the more cornering force a tire is able to generate.

A number of factors affect the coefficient of friction of a tire. The construction techniques used in a tire itself include contact patch area, bias angle, tire compound, the section height, aspect ratio and other factors. Vehicle factors will also affect the coefficient of friction, the most notable being the vertical load upon the tire contact patch. The surface of the race track will also have a tremendous affect on the coefficient of friction.

The goal is to maximize tire contact with the racing surface and maximize cornering force. To do this, we want to maintain the highest possible degree of traction. Traction has two components, the coefficient of friction and adhesion.

Adhesion is the force which holds together the unlike molecules of substances when two surfaces are in contact with one another. Unlike friction, adhesion is a molecular interaction or interlocking. Friction is the resistance to the motion of two moving objects on surfaces that touch. Four our purposes, the combination of friction and adhesion creates adhesive friction or traction.

The interlocking provided by adhesion allows us to exceed previously determined limits in straightline acceleration, braking and lateral acceleration or cornering force. This is due to the ability of racing tires to achieve a coefficient of friction greater than one. A coefficient of friction greater than one means that resistance to motion between one surface and another is greater than the load applied to that surface. In this case, it is the tire surface against the track surface. Obviously, a change in tire compound or in track surface will alter the coefficient of friction as well as the ability of the tire to interlock with the surface of the race track.

Load

Load is the total weight that is on the tire. At rest, this is equal to the weight on each corner of the car. In the dynamic state, the load shifts due to lateral and longitudinal weight transfer. As the load on a given tire is increased, the total adhesion on that tire increases. Conversely, as the load increases, the coefficient of friction decreases. Adhesion is the amount of traction a tire is creating. When you add load to a tire, you get more traction — up to a point. If you graph the curve of the increase, at lower G loading it is fairly linear (straight line). But as G forces increase, there is more of a curve and then a drop off. The heavier the car, the more rapid the drop off.

Loading versus cornering is a different story. A pair of tires (two front, or two rear) when evenly loaded, corner together better. The slip angle on a tire does not increase or decrease proportionally as vertical load is placed on the tires. And, more traction is available at a lower slip angle. So, during cornering, an outside tire loses traction quicker (because of more load from weight transfer) than the inside tire gains by a decreasing slip angle. The answer is to have a greater amount of static inside weight so during cornering the load on the left rear and right rear is exactly the same. This equalizes slip angles, thus optimizing traction.

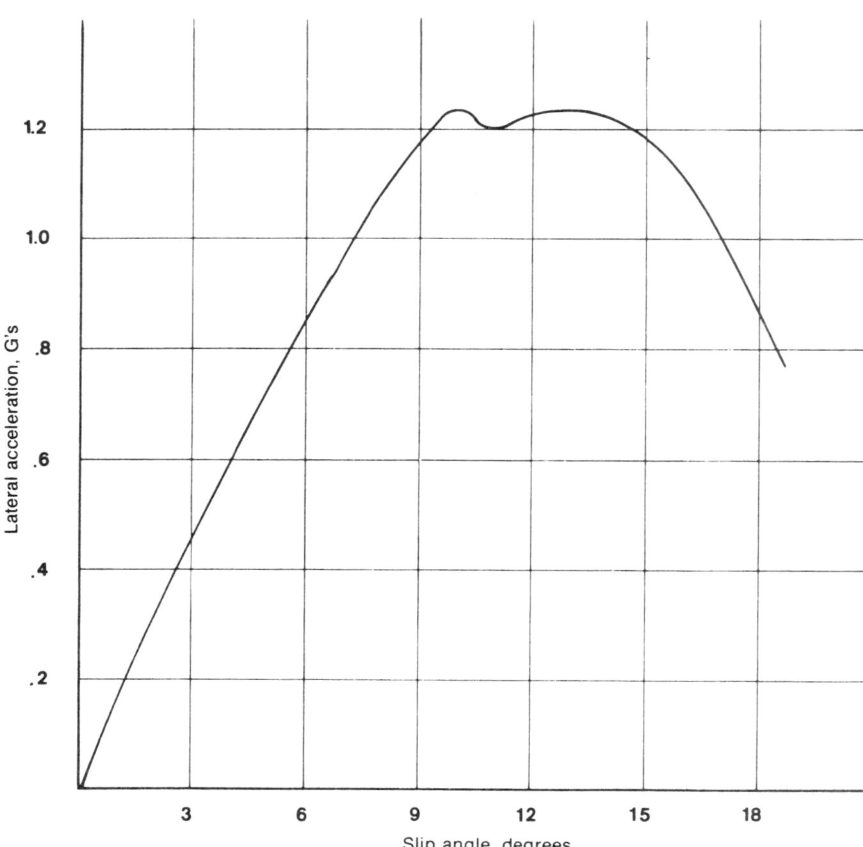

Slip angle vs. lateral acceleration. The actual cornering force is related to the slip angle of the tire in a similar way as the coefficient of friction is related to the slip angle of the tire.

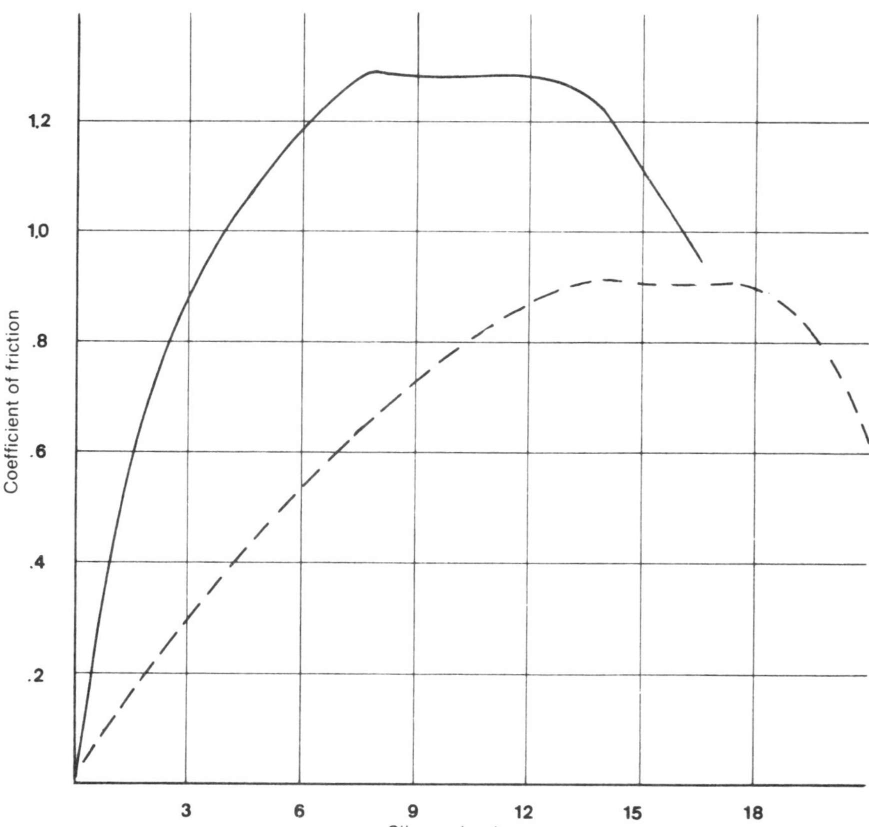

Slip angle vs. coefficient of friction. The solid line on the graph represents an asphalt racing tire while the dotted line represents a passenger car tire. Note that all tires will reach a maximum coefficient of friction, maintain it for a time and drop off quickly when the slip angle becomes too large.

Tire Slip Angle

To make an oversimplifcation, the slip angle of a tire is the angular difference between the wheel direction and the contact patch direction. This has nothing to do with the steering angle. Both front and rear tires operate at a given slip angle any time a lateral load is generated.

Design characteristics of tires that affect slip angle are compound, cord bias angle, cord material, aspect ratio (a higher ratio equals a lower slip angle), and sidewall height versus tread width (a higher sidewall or narrower tread width increases the slip angle).

Although an increase in vertical load will give us a net gain in cornering power, this will also increase the lateral load and can push the slip angle over the point where maximum coefficient of friction is achieved.

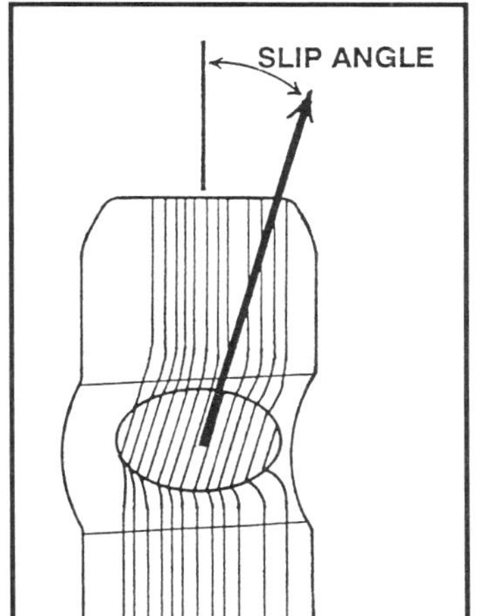

A number of factors will affect tire slip angles. Spring rates will affect the loading on the tire and thereby the slip angle. Tire pressures will affect tire slip angle, up to a point. Lower tire pressure will increase the slip angle at a given lateral force. Jacking weight will alter the tire slip angle by changing the loading.

The Traction Circle Theory

Tires really don't care in which direction they generate force — lateral, longitudinal, or any commbination of the two. This fact has a tremendous impact on how we drive and set-up a race car. The trick here is that a higher total force can be achieved by using the tire traction for two jobs at once, either cornering and braking, or cornering and acceleration. This phenomenon is best represented by a circle graph called the traction circle.

The traction circle refers to the fact that traction is non-directional. Tires will provide a given amount of traction at the limit of adhesion. Tires don't care which direction traction is being applied,as long as the limit is not exceeded.

If we represent this with a circle, the circumference of the circle is the limit of of traction. This limit can be shown as a radius of the circle drawn in any direction. A radius to 12 o'clock would show 100 percent of the available traction used for acceleration. A radius to 6 o'clock would be 100 percent braking. The 3 and 9 o'clock positions show 100 percent lateral acceleration (cornering) to the right or left. A radius falling any place else on the circle shows a combined force in two directions.

In other words, 1:30 would show half the available force used for acceleration, half for cornering to the right. 7:00 would be a third of the force for cornering to the left and two thirds of the force for braking. There are an infinite number of points on the circumference of the traction circle. This whole concept dictates modern driving techniques.

The diagram shows a traction circle plot and relates it to the path of a race car through a turn. From point A to point E, both braking and cornering are taking place, with braking forces reducing as E is approached. At F, all traction is being used for cornering. From G through K, the car is accelerating, with less and less traction used for cornering. At L, all traction is used for acceleration. From here until the entry to the next turn,the graph will drop straight down the vertical axis as aerodynamic drag increases, horsepower falls off, and braking begins.

Keep in mind that on a traction circle diagram, any point on the plot not falling on the vertical or horizontal axis represents a resultant force with both a vertical and a horizontal component that represent both lateral and straightline acceleration.

Tire Temperature Reading

Because all suspension adjustments and improvements are for the benefit of improving the bite of the tire on the track surface, it stands to reason that tire temperatures are the best indicator of what the tires and suspension system are doing. In fact, the tire temperature method is the only chassis tuning method where it is possible to get away from guessing and work with scientific accuracy.

The instrument that is used to measure tire temperatures is called a tire pyrometer. It is an electronic instrument which almost instantly gives a temperature reading when its probe is inserted into a tire surface.

While you may think that a tire pyrometer is an expensive tool, it is not in comparison to the investment made in a race car. Prices for quality tire pyrometers are about the same as just one racing tire.

Reading tire temperatures is the only way to sort a chassis on a paved track car, but it is also very effective for a dirt track chassis as well. Camber and toe-out can be very effectively sorted out on a dirt track chassis running on a skid pad.

Tire temperatures are read at three positions across the face of each tire and are recorded systematically on a sheet of paper in the manner shown in the accompanying examples.

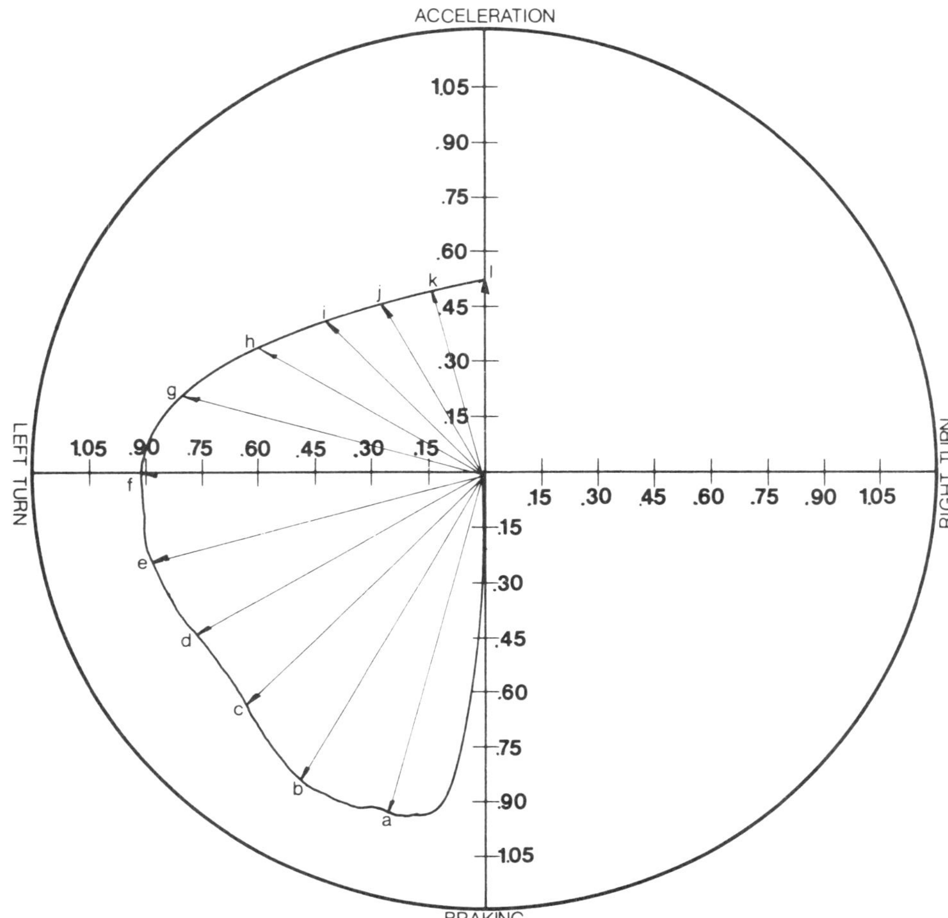

The arrows on this traction circle represent the various forces in G's being generated by the car in braking, cornering and accelerating. The solid line is a graphic representation of the forces on the traction capability of the vehicle. At 'A', the car is braking. At 'B', both cornering and braking are taking place. At 'C', 'D', and 'E', there is progressively less slowing, and more cornering. At 'F', all traction is being used for cornering. At 'F' through 'K', the car will be accelerating more, with less traction being used for cornering. At 'L', all of the traction available is being used for acceleration.

A vehicle's tires can develop a maximum amount of cornering force in any direction — accelerating, cornering, braking, or any combination of these. If we have a tire that is capable of generating 1.2 G's of tractive force, we can corner at 1.2 G's of lateral acceleration or we can brake at 1.2 G's deceleration. Or we can combine the two equally and have 1.2 G's of cornering and braking available. Understanding this concept will help the driver — especially on a paved track — how to better utilize the maximum capabilities of his tires and car.

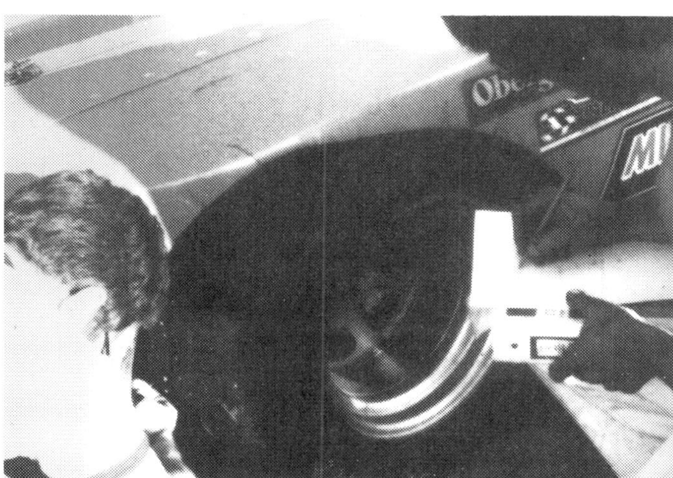

Tire temperatures can cool very rapidly. Take them just as soon as the car comes off the track.

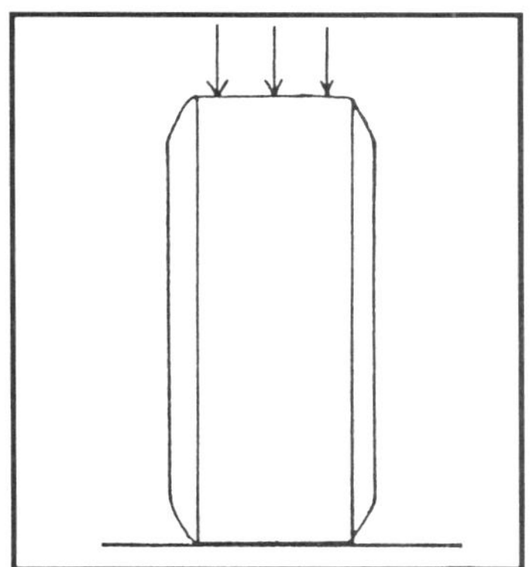

The tire pyrometer probe should be inserted across the tire face as indicated. The outside and inside edges should be measured about 1 inch in. Insert the probe at about a 45 degree angle to the face, not straight in.

By comparing the temperatures across the face of each front tire, it can be determined if each tire has too much or not enough negative camber, or if there is too much or too little toe-out, or if inflation pressure is correct.

Comparing the average temperature (of all three positions) of the right front tire to the average of the right rear will determine if the chassis is tending toward understeer or oversteer.

It is very critical to take your tire temperatures just as quickly as the car stops in your pit. The further the place you take the tire temperatures is located away from the track, the cooler the temperatures are going to be. And, the closer the temperatures will be to each other across the face of the tire as they tend to even out because of conductive heat transfer.

Tire temperatures can change very quickly and cool off very fast. Racing tires dissipate heat very quickly, and side to middle temperatures can equalize quickly. Take the tire temperatures very quickly, starting at the right front and working around the car in a circular motion to the right rear, ending up at the left front. Give the pyrometer needle time to warm up until the temperature stabilizes on the dial and then don't let it cool off when going from one tire to

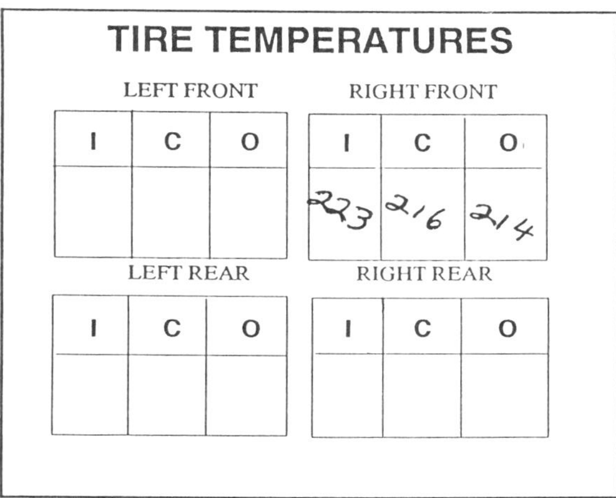

The more static initial camber, the greater the temperature difference between the outside and inside at the right front. Above is a good temperature spread on a 3/8 to 1/2 mile paved track with –2 degrees static camber. Below, the same application with –1 degree static camber. Second below, correct temperature spread for same application with –3 degrees static camber. The idea here is that the more static negative camber, the more inside heat at the right front tire.

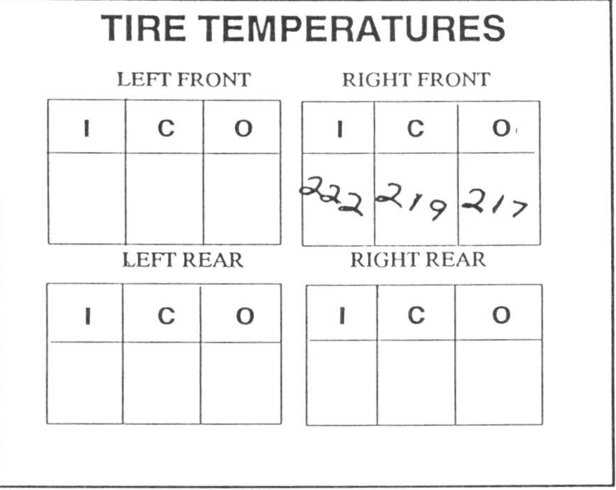

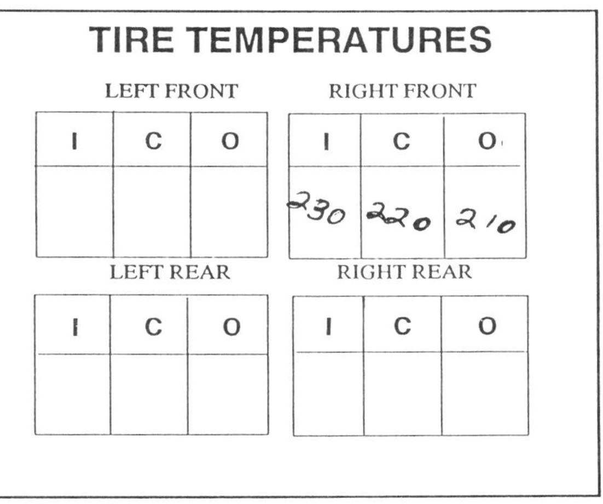

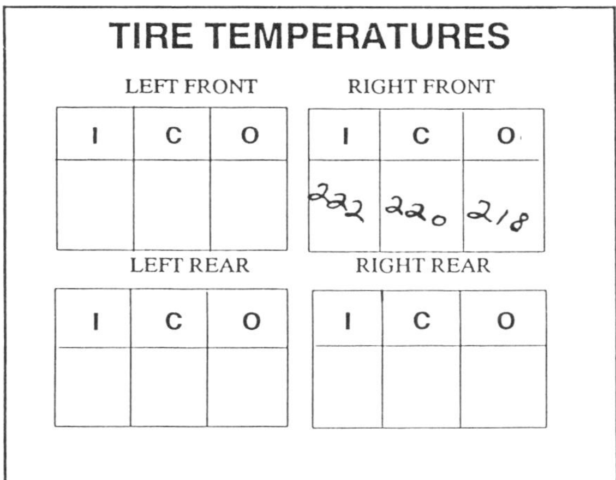

A short track, like a 1/4-mile, will show a more even heat pattern across the right front. Above is a good heat pattern for a 1/4-mile with –2 degrees static right front camber.

the next. This helps in getting a full set of temperatures quickly.

Don't measure the surface tire temperature, but rather push the pyrometer needle (gently — they're brittle) in at a 45 degree angle approximately 1/8 to 1/4-inch into the tire. When inserting the pyrometer needle, you want to go through the rubber surface of the tire and touch the cords of the body. This gives you the true and accurate temperature of the tire at the spot you are measuring. The inside and outside temperatures of the tire should be taken about one-inch in from the shoulder of the tire (see diagram).

In the following discussions, the terms inside and outside edges of the tires are used. The inside edge is always that one closest to the pole, and the outside is always closest to the grandstand.

Reading Camber

Perfect tire temperatures that reflect static negative camber settings should always show at least 5 degrees cooler on the outside for a 1/4 to 3/8-mile track. The longer the track distance, the greater the difference between the outside and inside edges of the tire. The more static initial negative camber, the greater the temperature difference. For example, if your car has 2 degrees of static negative camber, there should be 7 to 10 degrees of temperature spread from the outside to the inside edge of the tire (outside cooler). At 1 to 1-1/2 degrees of static negative camber, the temperature spread should be between 5 and 7 degrees. At 2-1/2 to 3 degrees of static negative camber, the temperature spread is going to be 15 to 20 degrees warmer on the inside edge.

The difference in temperature spread with varying amounts of static negative camber is the result of the way the tire goes down the straightaway. With static negative camber, the outside edge of the tire is off the track surface in varying amounts. The more negative static camber, the more tire surface that is off the track (or barely in contact with it) on the straightaway. This cools the outside edge of the tire.

The longer the track (or straightaways), the greater the difference between the outside and inside edge will be. On a short, tight track, like a 1/4-mile, the tire temperatures are going to be a lot more even across the face of the tire. This is because the corners are a greater percentage of the race track distance, and the outside edge has less time to cool on a real short straightaway. So, the shorter the track, the more even the tire temperatures are going to be, outside to inside (with a given amount of static negative camber); the longer the track (especially longer straightaways) the higher the inside edge temperature should be.

Don't ever try to get the tire heat even across the face of a tire. If you do, you will wind up with a push problem going in to the corner and in the middle of a turn.

Rear Tire Temperatures Versus Stagger

Stagger on the rear axle can actually be thought of a negative camber for the rear axle and wheels. So, with rear stagger present, you are going to see a temperature variation pattern on the rear tires. The inside edges will be hotter than the outside. The more stagger present, the more temperature variation.

Some Examples of Analyzing Tire Temperatures

The best way to understand the interpretation of tire temperatures is to read through some examples. Be sure to notice that references to the inside, in all cases, refer to the pole side of the car, and references to the outside refer to the grandstand side.

Example One

The right front is warmer on the outside edge than in its other two sections, indicating that it needs more static negative camber. The outside edge of the tire is warmer because it has been doing more of the tire's work than the rest of the tire surface, so it needs to be leaned in at the top.

The left front is too warm on its outside edge, so it needs more positive camber. That is, the top of the wheel needs to be leaned toward the inside (to the left as the driver sits in the car) so more of the tire's surface works on the track.

The right rear tire shows an average temperature of about five degrees warmer than the right front average. That would indicate this car is oversteering. Normally, an av-

TIRE TEMPERATURES

	LEFT FRONT			RIGHT FRONT	
I	C	O	I	C	O
190	195	210	215	220	225

	LEFT REAR			RIGHT REAR	
I	C	O	I	C	O
195	188	185	227	220	225

Example One

erage temperature on the right rear of 10 to 15 degrees cooler than the right front would indicate a neutral steering condition (no oversteer or understeer). As the two average temperatures approach the same number, the car is on the verge of oversteer. When the right front is more than 15 degrees warmer than the right rear, the car is understeering.

Also on the right rear a slight case of underinflation of the tire can be noticed. If after several hard laps of driving the center temperature is warmer than the other two, the tire is overinflated. In this case,a difference of about five degrees cooler in the center indicates underinflation. If the center temperature difference was only 1 or 2 degrees cooler, it would not make any difference (remember that the maximum cornering power can be increased with the least amount of inflation pressure you can use).

The temperatures on the left rear, with the inside edge warmer, indicates a certain amount of stagger in the car. If this car was running on a 3/8-mile to 1/2-mile paved track and had 55 to 57 percent cross weight in the chassis, these rear tire temperatures would be too cool. Large amounts of cross weight adds more heat into the left rear tire.

Example Two

Looking across the right front temperatures, the outside edge is much cooler than the inside edge (the inside is almost 20 percent higher), so a little more static positive camber is needed to tilt the top of the wheel outward.

With the camber problem corrected, the difference between the average right front and the average right rear temperature is too similar, meaning that the car is on the verge of oversteer. To make the right rear cooler, the front roll stiffness should be increased, either by increasing the spring rate of the front anti-roll bar or increasing the right front spring rate slightly.

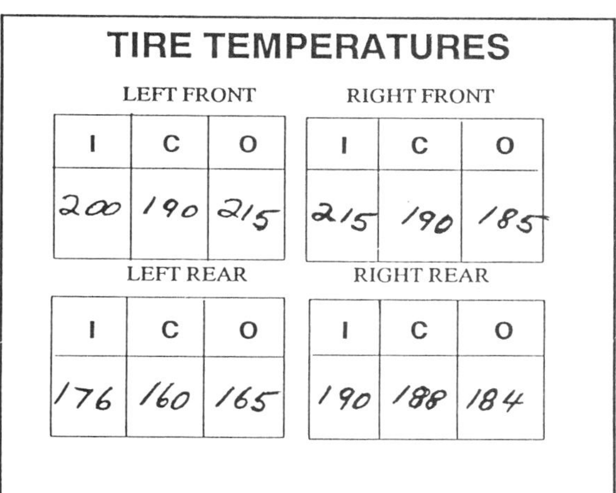

TIRE TEMPERATURES

	LEFT FRONT			RIGHT FRONT	
I	C	O	I	C	O
200	190	215	215	190	185

	LEFT REAR			RIGHT REAR	
I	C	O	I	C	O
176	160	165	190	188	184

Example two

The left front, with the inside edge some 15 degrees cooler than the outside edge, indicates it needs more static positive camber.

The left rear, with its cooler center temperature, could use a little more inflation pressure. The higher inside temperature indicates a normal stagger condition (there's only an 11 degrees spread from outside to inside — nothing to worry about).

The left rear has the coolest average temperature of any of the four tires on the car. If the cross weight was correct, the left rear temperatures would be warmer. The left rear needs some more weight on it to let it do more work.

Example Three

The toe-out indication is read on the inside edge of the right front and the outside edge of the left front. If the car has too much toe-out, the tires will heat up more in these two spots by the same amount of temperature excess. If

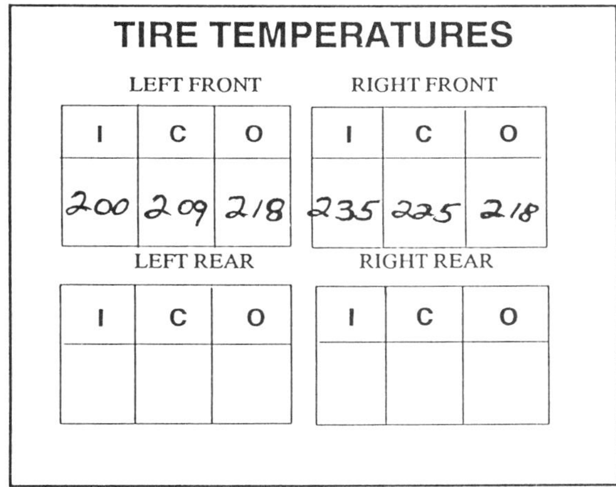

TIRE TEMPERATURES

	LEFT FRONT			RIGHT FRONT	
I	C	O	I	C	O
200	209	218	235	225	218

	LEFT REAR			RIGHT REAR	
I	C	O	I	C	O

Example three — too much toe out.

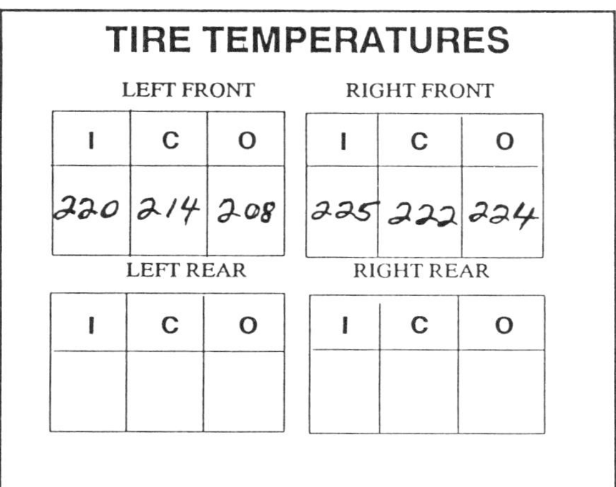

TIRE TEMPERATURES

LEFT FRONT			RIGHT FRONT		
I	C	O	I	C	O
220	214	208	225	222	224

LEFT REAR			RIGHT REAR		
I	C	O	I	C	O

Example three — too much toe in. If this car is running on a 3/8 to 1/2-mile medium banked track, the right front spread with ideal toe would be 225/220/215.

TIRE TEMPERATURES

LEFT FRONT			RIGHT FRONT		
I	C	O	I	C	O
194	196	199	225	220	217

LEFT REAR			RIGHT REAR		
I	C	O	I	C	O
215	210	205	205	198	195

Example four. Ideal tire temperatures for a 3/8 to 1/2-mile paved track, medium banking. Notice the average heat spread between the right front and right rear. Also notice the heat pattern across the right rear and left rear, indicating proper stagger.

there is too much toe-in, the outside edge of the right front and the inside edge of the left front will be too hot by about the same amount.

Example Four

This shows the ideal tire temperatures for a paved track car running on a 3/8-mile to 1/2-mile track, slight to medium banking. The average difference between the left front and right front tires is only about a 10 percent difference cooler for the left front. This indicates the left front is working. The heat pattern for the left front is slightly warmer on the outside. This shows there is more toe-out (Ackerman steering effect) at the left front, steering the inside turn radius more.

Left rear tire heat is good, showing enough cross weight to make that tire work properly.

The pattern of inside heat higher than outside on both the left rear and right rear shows the correct amount of stagger for proper rear tire "camber".

The patterns of heat going across the face of each tire from the inside to the outside, on all four tires, shows no drop or increase in the center position, which indicates there is no overinflation or underinflation of any tire.

The difference between the average of the right front and the average of the right rear shows almost exactly a 10 percent greater heat in the right front, indicating a neutral handling car. The right front average is a little more than 15 degrees warmer than the right rear, so the car is a little bit tight.

Problems That Cloud Tire Temperatures

Be sure that hard braking and cornering maneuvers from the last turn and getting off the track and into the pit do not influence the tire temperatures. If any of these have been used, be sure to allow for it. Also, it takes a minimum of 10 to 15 hard laps of running before a tire warms up to proper operating temperature and any chassis evaluations can be made. The shorter the track and the cooler the track temperature (or the harder the tire compound), the longer it takes for the tires to heat up. Don't do anything to artificially induce heat into the tires to bring them up to operating temperatures quickly — it will only mask the true chassis characteristics that the temperature patterns should show.

And, as we said before, take the tire temperatures quickly. When the tires are hot right off the racing surface, the small subtle differences will show up. When the tires have had a minute or so to cool, they are lost.

Orderly Steps To Interpret Tire Temperatures

All of the information provided in the above examples can be very confusing. Reading tire temperatures and getting the proper and correct interpretation can be an art. It takes time and practice to determine what works. To help you in the beginning to sort out a set of tire temperatures, we have the following set of priorities to look at to determine certain chassis handling characteristics:

1) Look at all four tires first to see if the is any overinflation or underinflation present (center temperature significantly hotter or cooler). If so, adjust tire pressures, and take another track run.

2) Look at the right front temperatures. Look for proper negative camber pattern.

3) Compare the right front average to the right rear average, looking for an indication of oversteer, negative steer or understeer.

4) Look at the right front outside to the left front inside temperatures for an indication of toe-out or toe-in. Be looking for an indication of either too much toe-in or too little toe-out.

5) Look for the proper camber pattern at the left front.

6) Look for an indication of Ackerman steering (more toe-out at that corner only) at the left front. The outside edge of the left front should be a little warmer.

7) Look for a proper stagger heat pattern at the left rear and right rear tires.

8) Look at the total heat of the left rear for an indication of proper cross weight.

9) Look for all tire temperatures to be within the normal operating temperature range as prescribed for the tire you are using. Generally, the proper operating range for a short track tire is between 190 and 230 degrees. Check with your tire dealer that you purchase tires from for the correct temperature parameters for the tire you are using. If there is a corner of the car that is too hot, a spring rate change is indicated. For example, if the average of the right front is 260 degrees, and the right rear average is 210 degrees, the right front spring rate must be decreased or the right rear spring rate must be increased.

10) If the average of all four tire temperatures is either too high (above 230 degrees) or too low (below 190 degrees), you might take a look at a tire compound change. If the average temperature is too low, the compound should be softer. If the average temperature is too high, the compound should be harder.

10) While you are looking at the tire temperatures and making interpretations, do not ask the driver for his evaluations of the handling. Let the tire temperatures give you your first impressions, then compare that with what the driver has to say.

11) Make chassis adjustments, according to what the tire temperatures indicate, and have the driver make another set of test laps. Continue to do this until the temperatures show a well handling car and the driver is satisfied with the feel of the car.

Continue Monitoring Tire Temperatures

Camber settings don't necessarily stay the same, even for the same track. So, tire temperatures should constantly be monitored. Track conditions and ambient air temperatures can change ultimate camber settings. And, if you pick up more speed, especially if you get into the turns harder, optimum camber settings will change. Keep on monitoring tire temperatures.

Your starting settings depend on the camber change curve built into your car's suspension. If your car gains

Reading the wear patterns of a tire can give meaningful feedback about the performance of the chassis and tires. The grain pattern on the inside edge of this right front indicates too much negative camber.

negative 1-3/4 degrees per inch of bump travel, and your buddy's identical car gains only negative 1 degree per inch bump travel, your initial camber setting at the right front is going to be less than his for the same track.

Reading Tire Surfaces

Reading racing tire tread surfaces is something of an art form. By reading the tread wear pattern on the surface of the tire tread, we are better able to determine handling characteristics of the vehicle. Combined with tire temperatures, we can make sound judgments about changes that need to be made to the suspension to improve handling characterisitics. Often the ability to read the tread wear patterns is more valuable than reading tire temperatures, especially on dirt tires. This is partly due to the characteristics of the race track where you are competing. As we have said, tire temperatures are more indicative of the last turn or two on a given race track. However, tread wear patterns will be indicative of what goes on all the way around a race track.

Normally, a tire's tread surface is a dull blackish gray color. If a somewhat shiny spot appears on the tread surface, this is normally an indication of that portion of the tire being overloaded at some point. If a portion of the tread surface shows no wear signs at all, this indicates the tire has no load on it in that area. Because of the characteristics of racing tires and the desirability of dynamic camber change, the inside edge of the tire will generally wear somewhat more than the rest of the tire. If the wear is excessive, it indicates too much negative camber. If the outside edge of the tire is wearing more than the inside edge or the center of the tire, this is usually indicative of too much positive camber or not enough negative camber.

If little bits of rubber are rolling up on the tread surface, this indicates the tire is running too hot. If the grain pattern is too smooth, it normally indicates the tire is not running hot enough. The higher the tire compound temperature, the lower the tear strength of the tire surface. In other words, you can get the tire surface overheated to the point that track surface abrasion will pull chunks of the tire surface off.

If the front tires show more wear or excessive wear relative to the rear tires, this indicates the vehicle is understeering. If the rear tires show greater wear, this indicates the vehicle is oversteering.

If the center of any tire shows excessive wear, it indicates too much inflation pressure.

Developing the ability to read tires is important. One method which is effective to develop this quality is to study the tires of your top competitors. Compare the wear patterns and grain patterns on competitors' tires relative to their performance on the track. By practicing this, you will become adapted to reading the tires on your own car and subsequently be able to improve handling and increase the performance of your own race car. Along with other records, you should maintain records of tread wear patterns and how they affect the vehicle on the race track.

Racing Tire Break-In Procedure

Racing tires are very influenced by heat the first time they are run. They go through a "final cure" in their first use heat cycle (first run on the race car). At this time the resins in the compound stabilize, the rubber cures, and the entire tire grows and assumes its final dimension. After this break-in final cure, the dimension of the tire is much more stable through the rest of its use. During the first heat cycle of the race tire, when it grows to its final dimension, it runs hotter than during its normal operating range.

Race tire break-in should involve gradually working the tire up to racing speed for a few laps until the tires reach normal operating temperature. They should not be run for more than a couple of laps at normal operating temperature, nor should they be abused or run hard the first time. If they are, they will change compound hardness or blister, and be ruined as a suitable racing tire.

The biggest mistake racers make is trying to get racing tires too hot too quickly. Racers use too low a tire pressure initially and ruin the tire. The tire should be broken in slowly, with proper inflation, running it only two to three laps at racing speed, and then allowing it to cool.

Tire pressure relief valves can take care of the pressure build-up problem. Maximum operating pressure can be adjusted.

Tire Pressure

Lowering the tire pressure will create a larger tire footprint on the track, but it will also create more tire heat. 4 to 6 PSI pressure can make a big difference. When the inflation pressure is too low, the tire will get hot very quickly. This occurs because the cords of the tire body have much more friction.

Tire pressures will vary according to the type of tire being used and track conditions. Ask the tire man you purchase your tire from for a recommendation for that particular tire, track, and car in question. And, on asphalt tracks, use a tire pyrometer to fine tune the pressure settings.

The total weight of the race car also has a bearing on the inflation pressures use. The lighter the car, the less pressure required. The heavier the car, the more needed. The reason is that the heavier the car, the more pressure needed to support the tire carcass and prevent deformation of it.

Tire pressure build-up is a problem to be aware of with paved track tires. One way to regulate the pressure build-up is to use a tire pressure relief valve. A pressure relief valve is a popoff valve which is mounted in the wheel and can be set for a maximum operating tire pressure. Once the tire reaches that pressure, the popoff valve opens and the excess air pressure is bled off. This valve can be very effective in asphalt track chassis tuning because a deviation of as much as even 2 PSI from the optimum tire pressure can make a difference in tire performance.

Tire Use And Aging

When a tire is used over and over again, it gets harder and loses its grip of the track surface. The tire has been heated and cooled so many times that the compound oils are sweated out and the tire gets hard. At this point, it is very inefficient to grip the track properly.

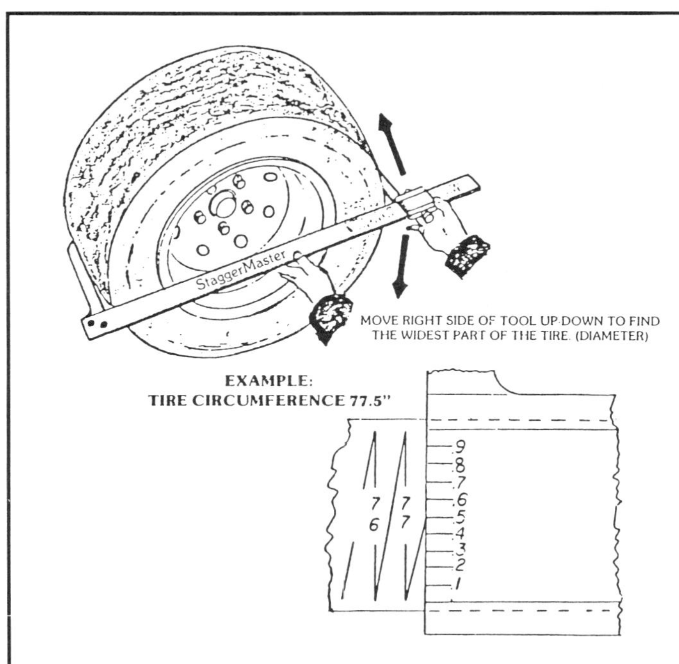

The Stagger Master gauge is much more accurate, and easier to use, than the old tape measure method.

Tire Stagger Tips

The shorter the wheelbase of the car, the more critical that stagger becomes to handling. As the wheelbase gets shorter, the car becomes less forgiving of improper stagger.

High tire operating temperatures affect tire stagger more. If the handling is wrong and the tire temperature gets too high, the tire size will grow more quickly and tire stagger will get out of hand.

The most effective way to measure tire stagger is to use a Stagger Master gauge. When using it to measure a brand new tire, be sure to take measurements in several places around the tire. This is because most new "sticker" tires aren't perfectly round, and numbers taken at different places can vary. It is best to scuff your tires first, which will make them round, before working with stagger.

Dirt Track Tires

All things discussed for tires in general in this chapter apply to dirt track tires. But one thing sets dirt tires apart from all others — grooves. Grooves run both laterally and radially and thus form blocks. The blocks create track cutting/gripping edges which produce more bite on a dirt surface.

The leading edges of the blocks are the critical parts which cut and dig in to the dirt surface. As a dirt tire wears, these leading edges become rounded off. There are two ways of restoring the sharp gripping edges — reverse the running direction of the tire on the rim, or regrooving the tire.

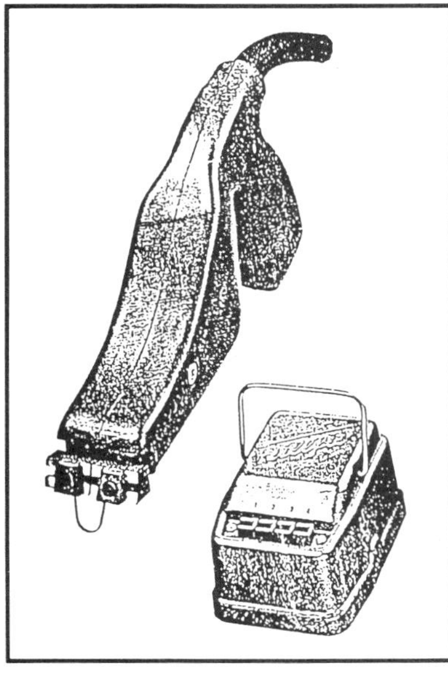

This tire grooving tool for dirt track tires is available from Fox Racing.

Reversing the direction of the tire will give it new sharp leading edges on the blocks to grip the track. However, if the tire is too worn, reversing its direction of travel can be more detrimental than helpful.

The leading edges of the blocks can be regrooved with a grooving iron to sharpen them again. Be aware that tire regrooving is a real art — let an expert do it for you, or teach you to do it. The grooving iron can also be used to create additional grooves on the tire if the tire is running too hot. Creating smaller blocks make the tire run cooler. This is created by more air flow around the greater number of blocks which aids in cooling.

Another trick to add traction to a dirt tire is "siping." A siping tool (which uses two sharp blades placed about 1/4-inch apart) is used to create more track gripping edges across each row of blocks on the tire. Tire siping is most helpful on a dry slick track where tire traction is very delicate when the tires are cold. The sipe cut should be not much more than just a scratch on the block surface. All it has to do is add extra grip for a few laps until the tire comes up to operating temperature. If the sipe cuts are too deep, however, they will fill with dirt before they wear off and cause the tire to skate across the track surface.

Working With Street Tires

At some tracks using Pro Stock chassis rules, street type tires must be used instead of racing tires. You cannot race competitively with a treaded street type tire with the tread at stock depth. Tread thickness is a very important characteristic for a racing tire. For tires engineered specifically for racing, this factor has been considered. But with a street tire used for racing, you have to become your own tire en-

Street tires must have the tread face depth shaved to 4/32-inch for proper performance. Some racing tire makers already make their D.O.T. tires to this spec.

gineer. The stock depth will store excess heat. The thicker the tread depth, the more heat build-up. This excess stored heat will eventually cause degeneration of the rubber compound, and eventually separation of the tread rubber from the cord and tire body. Heat is also generated by tread deformation or "tread squirm," which is also a very upsetting handling influence. So, what this means is if you have to race on street tires, you have to shave off the tread depth to at least 50 percent of new depth. Normal new tire tread depth is 9/32-inch. The new street tire should be shaved to 4/32-inch for racing. Tire retreading shops can handle

EVERY time before you go out on the race track, double check the torque on the wheel lug nuts. Use a torque wrench for more reliable results.

the tread face shaving for you. Some companies, such as Hoosier and McCreary, manufacture a D.O.T. "street" legal tire for racing which already has a reduced skid depth so racers don't have to have the tire tread shaved.

Racing With Radial Tires

Racing radial tires, because of their style of construction, require changes in the chassis. Radial tires need a lot of initial negative camber setting (camber thrust) at initial turn-in to a corner. A camber change curve, even with a radical change, just isn't enough to get it done. You need the negative camber instantly.

Radial tire sidewalls are very stiff. That changes everything. Soft, pliable bias ply tire sidewalls are forgiving, and add suspension and driver forgiveness. Radials do not. Radials require more driver sensitivity and smoothness. The stiffness of the radials offer some benefits, though too. Radials run cooler. There is less tread and sidewall movement so the tires stay cooler. This helps them stay more consistant. And because of their type of case construction, they do not change in diameter — thus no change in stagger. Radials are also very puncture resistant.

Rim Size

Keeping the rim width sized properly to the tread face width is very important for good handling. If you put a wide tire on a narrow rim, you get excessive leverage on the tire sidewall from the rim. This distorts the tire face, pulling the inside edge off the track. If you end up with a stiuation such as this, however, you will need a much greater amount of negative camber change to compensate for tire distortion.

Tire/Wheel Preventive Maintenance

Be sure to use rubber valve stems, not steel. If a race car has contact with another car or encounters debris, a rubber valve stem is more likely to flex while a steel one will break.

A good deterrent to tire leaks caused by debris penetration to the tire is the use of Marsh Racing Tires' Urethane Tire Sealant. The sealant is used to coat the inside of a racing tire, and when dry, it forms a tough air-tight film which seals air leaks. The stuff is a little expensive, but is very cheap insurance if you consider the alternatives.

New wheels should be tested for air leaks by installing a tire, inflating it and immersing the assembly in a dunk tank to look for leaks.

Bent wheels are a source of air leaks. If you suspect a leak with a possibly bent wheel, install a tire, inflate it, then paint the rim/tire joint with a liberal dose of 10% dishwashing liquid detergent/90%water, and look for bubbling.

A tire bead sealer should be used on a tire when mounting it to a wheel to help insure proper bead/wheel mating, a place where considerable tire leaks occur.

Chapter 9

Chassis Adjustment

One of the major problems of the Pro Stock late model sportsman car is that it is a very underpowered car — a low power-to-weight ratio. So chassis set-up is very critical. A more highly powered car has the torque and horsepower to overcome some handling deficiencies, but the 200 to 250 less horsepower in a Pro Stock car makes proper chassis set-up extremely important. You can't afford to have front tires that point the wrong direction in a turn, or a rear end that tracks at an angle — or anything that causes a drag on one or more tires on the car. Camber, toe and steering has to be right, and the rear end has to be square to the chassis as the body rolls in order to keep the tire contact patches from scrubbing off speed. The car that has it all sorted out is going to have the advantage.

Proper chassis adjustment will prevent having a car that won't go where you want it to and is all over the track.

Check for all possible chassis binds. The evidence shows how much turning radius can be required. Note how the shock is in the way.

Checking For Chassis Binds

Before doing your chassis set-up in the shop, you must look for the presence of any chassis binds in your completed car. Move each of the wheels through at least two inches more than their normal wheel travel. Carefully observe the movement of everything attached to that wheel. Look for shock absorber binding or bottoming out, A-arms moving freely or contacting the frame, the steering shaft moving freely when turned without contacting anything, a free movement of all steering components through full range of left-to-right steering with no binding or contacting, the Panhard bar moving freely with no binds or without contacting any chassis parts, and the rear suspension arms moving freely with no binds.

If you observe any problems, be sure to correct them right away before proceeding to the chassis set-up.

Squaring The Rear End

Simply, the rear end squaring process is making sure the rear end housing is set straight in the car — perpendicular to the vehicle centerline and not angled. If the right rear is set behind the left rear, the car will be loose. If the right rear is set ahead of the left rear, it will push.

Squaring the rear end is very critical to the car's handling. Even a 0.25-inch out-of-square can have a significant effect on handling.

Most chassis manufactured by professional builders have built-in squaring reference marks on the frame rails. These

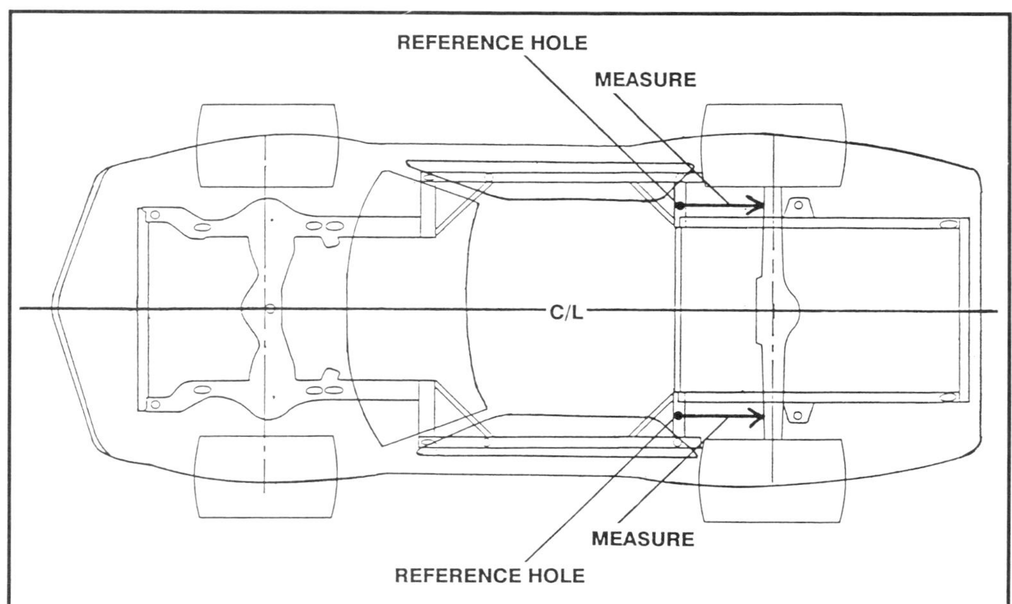

A very reliable reference mark for squaring the rear end is the jig locating hole on a profesionally built chassis (arrow). If you build your own chassis, drill parallel holes for reference.

references are usually holes drilled in the frame parallel to each other. Make sure you know where these are. If you have built your own chassis, make sure you remember to include these squaring references on the frame rails.

If your chassis is equipped with squaring refence marks on the frame rails, squaring the rear end is very easy. Simply measure straight back to the rear end housing at the same point on each side from the reference marks. The measurements should be identical. If not, the trailing link lengths must be adjusted to make them equal. If you car is not equipped with the rear squaring reference marks, you should add them. This is by far the quickest and easiest method of squaring the rear end. There will probably come a time when you will have to do it in a hurry at the race track, and you will be glad you added the marks.

The other way of squaring the rear end is commonly called "stringing" the car. It is done by stretching string tightly between jack stands on each side of the car running parallel to the wheelbase to establish outside reference lines. Measure from the string at several points to the frame rails to make sure the string reference line is absolutely parallel to the rails, and that each string is par'llel to the other. Then measure from the string to the front side and the rear side of each rear tire (see drawing). Make sure the measurement hits the rear tire in the same point on each side. These measurements should be identical. If they are not, the rear end housing is not square.

If the front measurement on the right side is greater than the rear one, it indicates the rear end is pulled ahead on the right side. Lengthen the right side trailing link slightly, and take new measurements.

If a string is stretched between the two rear jack stands running parallel to the rear end housing, measurements can be made from it forward to the rear end housing on each side to double check the squareness.

Setting Ride Height

The ride height should be set on the finished chassis in the shop before the weight distribution is set. Ride height is measured from the flats of the bottom of the frame rails at the forward and rearward corners of each side of each rail (see drawing). Once ride height is set, the front/rear and left/right weight distribution can be set on scales with ballast. Using ballast to set the desired weight distribution will have very little effect on the corner heights.

For dirt track cars, use the following corner heights:

Left Front — 4.5"   Right Front — 5"
Left Rear — 5.5"   Right Rear — 6"

For asphalt track cars, set the chassis as low as your sanctioning association rules will allow (be sure the chassis is designed for the ride height you intend to use. The numbers we quote below were design numbers we used

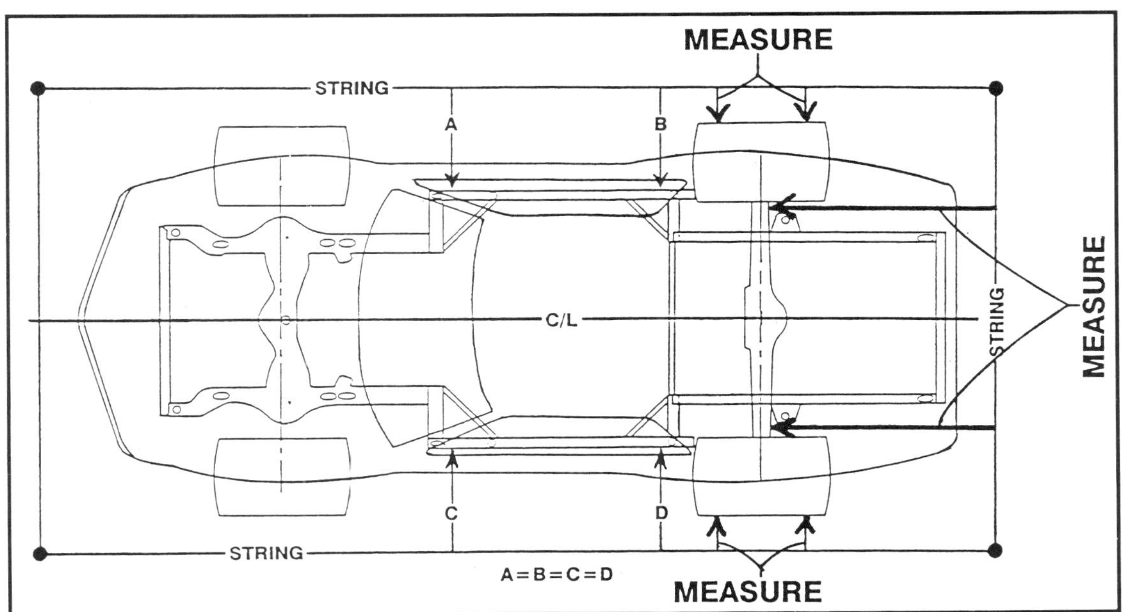

Stringing the chassis

with our paved track chassis blueprint.) Assuming the minimum ground clearance is four inches, use the following corner heights:

Left Front — 4" Right Front — 4.5"
Left Rear — 5" Right Rear — 5.5"

Dirt track cars need more ride height (ground clearance) than asphalt track cars to improve side bite (which is cuased by more overturning moment and body roll). For cars running on dirt tracks that get very hard, dry and slick, the ride height should be the same as quoted for asphalt track cars. For very loose and wet dirt tracks, the ride height needs to be slightly higher than the numbers quoted to help induce more side bite.

Ride Height Setting Short Cut

Once your car's corner heights and corner weights are set, it is a good time to cut and fit clearance blocks which can help you quickly set up your chassis without the need for corner scales, a flat surface and measurements at each corner. When each corner height and weight is finally set and right, there exists two relationships which will always remain true: the angle of the upper A-arms and the distance between the frame and the lower A-arm (at the front), and the distance between the top of the rear end housing and the frame at each corner of the rear. Knowing that these space relationships will always remain the same, small blocks of wood or metal (clearance blocks) can be cut which fit right between these reference surfaces. Once the car is perfectly set, fit and trim these blocks for each corner of the car, and record exact angle measurements for each of the upper A-arms. These can become your quick set-up tools whether you are at your shop or in the pits on a very uneven surface and need to reset the car.

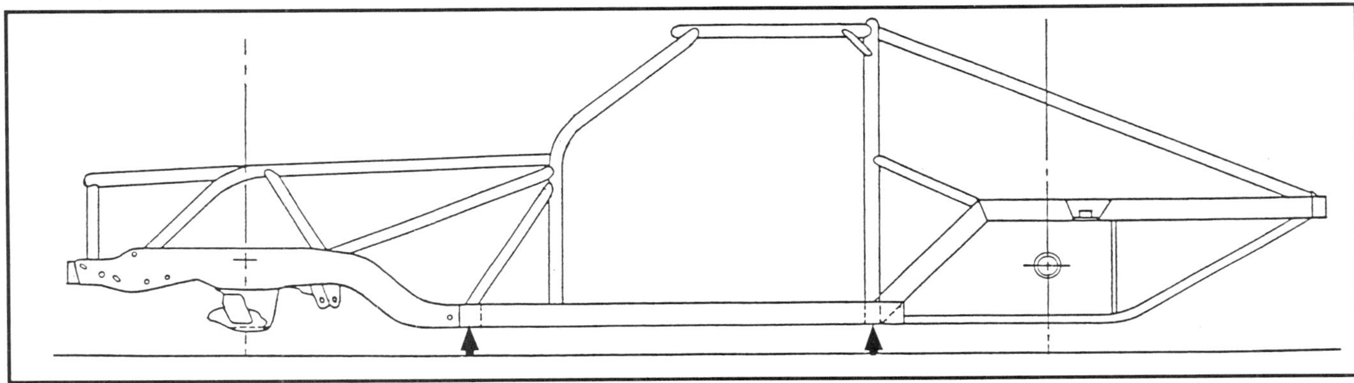

Measure for ground clearance at the forward and rearward corners of the flats of the frame rails.

Clearance blocks, such as seen to the right of the tape measure, will fit between reference surfaces of your chassis to help speed ride height setting.

Rear Tire Stagger

Stagger is the difference in inches of the tire circumference between the left rear and right rear tires. When the right rear is larger in circumference than the left rear, we have stagger. When the left rear is larger in circumference than the right rear, we have reverse stagger.

The minimum stagger required will vary from car to car and track to track. The variables include car track width (left rear tire center line to right rear tire center line) and the turn radius of the race track. Minimum stagger is the difference in size of the right rear tire from the left rear tire, determined by these variables. Because the two tires are running on different radii, one must travel further than the other (the outside tire must travel a further distance on a wider arc). This is accomplished by the outside tire being

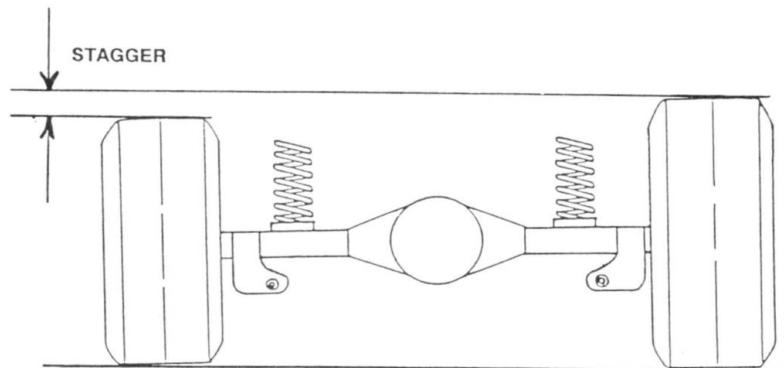

Tire stagger will help cure a push problem, especially on wet dirt such as this.

larger in diameter than the inside tire so it runs at a slightly faster speed.

Stagger helps get a car into a turn without pushing. And, it helps to overcome the pushing tendency of a car as it changes direction from straight line running.

Tire stagger gets a car off the corners quicker, which means more straightaway speed. Improved corner speed with more straightaway speed helps you set up passes more easily. Does stagger hinder straight line speed? Yes — it's a dragging force. But more acceleration off a turn more than offsets this. You need enough stagger to launch the car off the turns to gain more straight line speed.

To find the minimum stagger your car needs, use the formula:

$$\frac{D + .5\ (TW)}{D - .5\ (TW)} \times C_L$$

where **D** is the track turn diameter in feet, **TW** is the rear track width of the car in feet (divide the inch measurement by 12), and C_L is the left rear tire circumference in inches. The answer will be the minimum circumference of the right rear tire.

From this you see that the tighter the turn, the more stagger the car needs. If a car pushes in the middle of the turn and beyond, more stagger is required. If the car is loose, decrease the stagger.

Street-type tires require more stagger, but it is harder to come by. By the design of the tire, street-type tires are not as sensitive to circumference change with air pressure increases. The best thing to do is measure different tires to find the stagger you want.

An important concept to remember with rear stagger is that it is also rear camber on a straight axle. If you put a smaller circumference tire on the left rear, it puts negative camber in the right rear and positive camber in the left rear. This helps put the tire footprint flat on the track during cornering.

The key to going fast at any race track is keeping the momentum up through the middle of the corner. Stagger will help keep the car freed up for more momentum.

The ultimate guideline for rear stagger is to use as much as is necessary to get the car to roll easily around the middle of the corner and not get loose on corner exit. Tire temperatures should also be monitored closely to be sure they stay within reason across the right rear (inside edge of the right rear is going to be hotter, showing a tire temperature spread very similar to the right front with proper camber).

However, if you resort to large amounts of stagger to get the car through the turns, you are going to sacrifice turn exit traction plus increase tire wear.

Air pressure is going to effect tire stagger more than anything. For the feature race set-up, you must know what the increasing tire pressure build-up is going to do to your set-up. If you know your car handles best at 2.5 inches of stagger, but it increases to 4 inches through a feature race, the car will get so loose all you can do is hang on. You won't be able to race anybody. On the other hand, if the car loses stagger, like going from 2.5 to 1.5 inches, the car is going to push so much you won't be competitive. Be sure to measure tire growth and air pressure build-up over a set number of laps so you know where to start so that you end up with the ideal set-up at the closing stages of a race.

Where you want your set-up at the end of a race depends on the track surface history and the length of a race. Generally, you'll want to be tighter (pushier) for a longer race and a track surface known to get slick or slippery.

The key to going fast at any race track — dirt or asphalt — is to keep the momentum up through the middle of the corner. You are running on two different radii through a turn, and stagger is what will make a car run on the different radii. This will keep the car freed up so it will roll around a corner instead of having to be "driven" around the corner. This will allow you to pick the throttle up without getting the rear end loose or creating a push.

Sometimes the amount of stagger used ends up being a compromise, depending on the length of the straightaways. For example, with long straightaways, too much stagger is going to make the car move around on the straights and get a little loose on corner entry, even though the car needs the stagger for a tight turn. The answer is a compromise between what is the fastest **and** what makes the driver the most comfortable.

Stagger versus crossweight is also a very important consideration. A car with a high amount of crossweight is affected more by stagger than a car with less crossweight. A car with a very heavily loaded left rear wheel is more critical for stagger. If it is off some, it will really be detrimental to handling.

Basic Guidelines For Using Stagger

1) More stagger makes a car looser coming off a turn and heading into a turn.

2) Less stagger tightens up the rear of the car.

3) If a car pushes coming off a corner, add more stagger.

4) The less rear weight percentage a car has, the less stagger the car needs.

5) The tighter the turns (shorter turn radius), the more stagger required. For example, the minimum stagger required with a 64-inch rear track on a tight 1/4-mile track might be 4 inches, but the same car on a wide sweeping 3/4-mile track (with the same banking) might be only 1.2 inches.

6) A wet,heavy dirt track will use much more stagger to help turn the car — maybe in the neighborhood of 4 to 5 inches with a 64-inch rear track car on a 1/2-mile track.

7) A dry, slick dirt track will use much less stagger. For example, with the same car and track as noted in the above example, the dry/slick track would only require 1 to 1.5 inches of stagger.

8) More crossweight requires more stagger to get the car into and out of the turns. Conversely, less crossweight requires less stagger.

A wet, heavy dirt track will require more stagger to help turn the car.

A thick, heavy tape measure is the wrong tool to measure stagger. Use a very thin, flexible tape.

To get an indication of how much a tire will grow after it is run, inflate a right side tire to 60 PSI and let it set for 20 minutes, then bleed it down to 30 PSI and remeasure.

9) When picking out unmounted tires, keep in mind that the tires will grow about .75 to 1.0-inch after they are mounted.

10) To get a good indication of how a tire will grow in circumference when run, mount the tire and increase the right side tires to 60 PSI and hold it there for 20 minutes (first be sure that it is safe to increase the tire pressure that high). Then bleed the tires down to 30 PSI and immediately take a circumference measurement. That number should give you a good indication of the final tire dimension after they are run. Mark the sidewalls with the circumference measurement. Different bias ply racing tires will grow at different rates, even though they are theoretically produced equally. (Radial tires are a completely different story. Because of their type of construction, they produce very little circumference difference between different tires.)

11) For an indication of tire growth of left side tires, follow the above procedures, but inflate the tires to only 50

After scuffing the tires and bringing them up to operating temperature, remeasure the stagger.

PSI, hold for 20 minutes, then deflate them to 20 PSI and measure the circumference.

Mass Placement

Mass placement is where major concentrations of weight are set in the car. Concentrated weight masses in the car include the engine, ballast, fuel cell, battery, etc. This placement is in relationship to the location of the roll centers, roll axis, and CGH. Mass placement interreacting with these make a difference on how the chassis reacts with the roll of a car into a corner. Understanding mass placement is very important in understanding how the entire chassis works together because it goes hand-in-hand with roll centers and spring rates.

A discussion of this concept is best described by an example. A racer with a 3,200-pound car (55 percent left, 55 percent cross, 50 percent rear) running on a high banked paved track had a problem getting into the turns. Instead of following the groove and turning down into the corner on entry, the car would climb up the bank and push up in the middle of the turns. Tire temperatures showed the left front to be 20 to 30 degrees (average) cooler than the right front. That indicated the left front wasn't doing very much for the chassis.

The first rection of the driver of the car was to put a stiffer left front spring in the car, but if you think through what would happen during body roll, the stiffer spring would not be an effective change because it would not have enough mass to react against. The answer was to get enough mass at that corner.

The car had lead ballast bars bolted onto the frame rails on each side — a 40-pound bar on the left side at the rear of the frame rail, and a 40-pound bar on the right side at

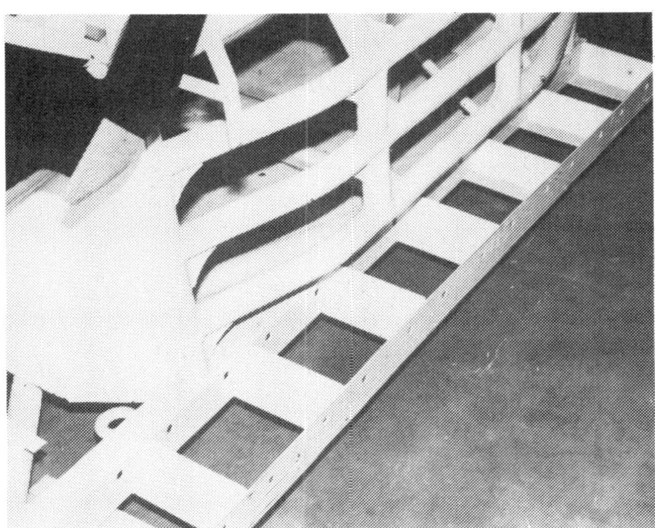

A weight tray allows mass to be shifted front to rear, etc. An open frame rail with ballast inside it will work too.

the front of the frame rail near the intersection with the front clip.

To adjust the mass placement, the chassis engineer observing the test reversed the position of the two lead bars. He moved the left side to the front, and the right side to the rear. Then the car was reset with the same starting crossweight and corner weights and heights as before.

The car then would turn down into the corners. No change in springs, or roll centers, was required — just a change of mass placement.

The whole problem was that both front tire contact patches weren't being used to turn the car down into the corners. Only the right front was, and tire temperatures indicated a lack of adequate use of the left front tire. When the mass placement was changed, the down loading on the mass put more weight on the left front tire — even though the crossweight stayed the same.

This same mass placement theory is valid for dirt cars as well. For example, a racer with a dirt track car that was designed with mass placement for the engine to be offset two inches to the left moved the engine back on car centerline. The result was that the car carried the left front tire up through the turns. There was not enough mass placement at the left front corner of the car. The racer's first adjustment, however, was to stiffen the right rear corner spring rate. This didn't cure the problem, and in fact it added to the chassis problems by loosening up the rear of the car on turn entry. The answer here, again, was to get more mass placement at the left front corner. If he didn't want to (or couldn't) move the engine back to its left offset position, ballast could be added to the left front frame rail to get the mass placement balanced once again.

The idea with mass placement is to get the weight placed in the chassis so that each tire contact patch is used to its ultimate during cornering. Build the chassis with all the components in place where they need to be, and keep the total vehicle weight in mind. Try to minimize total weight every place you can (but don't ever sacrifice safety). If everything is done correctly, you should have to add between 400 and 500 pounds of ballast to the car to bring it up to legal racing weight. Then add ballast to the chassis in the places needed for proper weight distribution.

Read through the following "Chassis Set-Up At The Shop" section to see how this is done.

Suggested Weight Distribution

Dirt Tracks

For dirt track cars on all but high banked tracks, use the following weight distribution as a good starting "ballpark":

Left side — 53 percent
Rear — 53 percent
Cross weight — 49 to 50 percent

On most successful dirt cars, the cross weight will fall right in between 49 and 50 percent, usually at about 49.7 percent.

If you go past 53 percent left side weight percentage on a flat to moderately banked dirt track, you are going to have trouble getting side bite at the right rear, and the car will have a tendency to be loose on corner entry. More than 53 percent rear weight will have a tendency to cause a push going into a turn and coming out.

If you are running on a high banked dirt track, a higher left weight percentage (55 to 56 percent) becomes more beneficial because the car doesn't need the side bite at the right rear. The bite will come from the down force created by the banking angle.

Paved Tracks

For paved track cars running on flat to moderately banked tracks, use the following weight distribution as a good starting "ballpark":

Left side — 55 percent
Rear — 51 percent
Cross weight — 55 percent

Many racers discuss the use of 58 or 59 percent left weight percentage on paved track cars. But with the type of chassis required with a Pro Stock class of car (stock front clip, no offset, perimeter frame) it is very difficult to get much more than 55 percent left weight, even if the car is built light enough to require 300 to 400 pounds of ballast to be added. That is why we have chosen the above numbers as a very sensible and achievable guideline.

Cross Weight

Cross weight — or diagonal weight, as it is also called — is simply the total of the right front and left rear corner weights divided by the car's total weight. More cross weight

Using individual wheel scales is the only way to properly set the corner weights on a race car. Grain scales such as above are accurate, but cumbersome. The best type of scale is the electronic digital scales, such as seen below.

adds more bite or understeer into a chassis. It is used, mostly on paved tracks or dry/slick dirt tracks, to keep the rear end of the car tight on corner entry, and to improve bite off the corners. It favors the left rear tire contact patch by more heavily loading it.

To compute the cross weight, add together the single wheel weights of the left rear and right front corners. Then divide this number by the total vehicle weight. Your answer will be a percentage, which is the cross weight percentage. Cross weight relates to the same amount of percentage of weight set diagonally in the car, no matter what total weight it is, be it a 2,400 or a 3,200-pound car. You will want to work with your car to find the diagonal weight that works best with it, such as 51 percent, or 55, percent, etc.. Then, if you change the total weight of the car, you can still come back to that ideal bite set-up.

Cross weight works very closely with tire stagger. While cross weight helps put bite in the car for corner entry and exit, sometimes it will cause a car to push or understeer. At this point, more tire stagger is required to help balance the chassis. As a general rule of thumb, the more cross weight a car has, the more stagger it needs.

Chassis Set-Up At The Shop

Your goal in setting up the chassis at the shop is to have the car ready to race competitively as soon as it rolls off the trailer at the track. With the car set up properly at the shop, you should have to make a very minimum amount of adjustments at the track.

Be sure that you choose a flat, level surface in your shop on which to do the set-up. Always use the same place. Make marks on the floor where the car sets so it can be returned to the same location time after time.

The chassis set-up should follow a specific order each time. Make sure the correct springs are in the car, and that the car is completely race-ready. All fluids should be full, and the wheels and tires (including tire stagger) and air pressure should be the same as you plan to use at the track. And, add the correct amount of ballast in the driver's seat to simulate the weight of the driver being on board. Only by taking all these steps can you guarantee a meaningful and consistant chassis set-up.

The order of chassis adjustment procedures should be:

1) Set ride height at each corner
2) Set the desired weight distribution
3) Set cross weight
4) Set caster
5) Set camber
6) Set toe-out

Setting the ride height is first, and that has been discussed previously in this chapter. However, for doing the initial weight distribution set-up, set the ride height at the right front and left rear corners 0.5-inch LOWER than the specified heights in the ride height section. This will help to get corner weights and cross weight correct along with the corner heights after the cross weight is set (this will become clearer later when we do the weight distribution step-by-step).

The next step is to raise the car up in place and put a wheel scale under each wheel. Be sure to use a known quantity of weight, first, to calibrate the scales and insure their accuracy. Once the car is in place on them, disconnect the shock absorbers and anti-roll bar to be sure they do not hold the chassis up off the springs.

When setting the corner weights of your race car, use only good quality wheel scales. You say you can't afford good wheel scales? Then pay a friend $5 or $10 to borrow his for set-up. Or go to a local chassis builder who has scales and pay him to use his. Or get together with three other racers and each of you can buy one wheel scale. When buying wheel scales, remember that if you cut corners on price and quality, you are cutting corners on accuracy too.

Once the actual weight and the target weight of the car is known, start adding ballast bars a little at a time. Keep checking the wheels scales to see how you are progressing toward the desired corner weights.

Read the wheel weights of your car and record them on a grid pattern on your paper for ease of visualizing the corners. Let's say that your finished race-ready car has to weigh 3,200 pounds with driver on board, and you built it light, so the initial wheel weights are:

725	720
640	635

The total weight of the finished car is 2,720 pounds. If it needs to weigh 3,200, 480 pounds of ballast will have to be added. The total left side weight is currently 1,365, or 50.2 percent. The total rear weight is 1,275 pounds or 46.9 percent. For a paved track application, the left side needs to be 55 percent, or 1,760 pounds total, and the rear has to be 51 percent, or 1,632 pounds. So, it can be seen that a good proportion of the added ballast weight will have to go toward the left rear corner and the left side. Start by adding ballast to the chassis a little at a time, and watch what happens to the weights at each corner. Be sure to keep the ballast as low as possible, and make sure that all ballast, when finally set in place, is SECURELY FASTENED to the chassis. Keep on adding ballast, and repositioning it, until the target wheel weights are achieved.

The target corner weights for a 3,200-pound car on a paved track, with 55 percent left and 51 percent rear weight, would be:

862	706
898	734

If you add up the numbers, you will find that we have achieved 55 percent left weight and 51 percent rear weight, but the cross weight (right front and left rear added together and divided by the total weight) is only 50.1 percent. The cross weight has to be set by screwing down on the right front and left rear weight jackers. It cannot be obtained with ballast. And, you will find that no matter what cross weight you use, the left side will always total the same, and the rear will always total the same. Ballast sets the left-to-right and front-to-rear weights. The weight jackers affect the diagonal weights. When the right front and left rear weight jackers are screwed down to set the cross weight, this will also raise the corner heights at these two corners.

Left, placing ballast effects left-to-right percentage and front-to-rear weight percentages. Right, jacking weight from the weight jack screws effects cross weight.

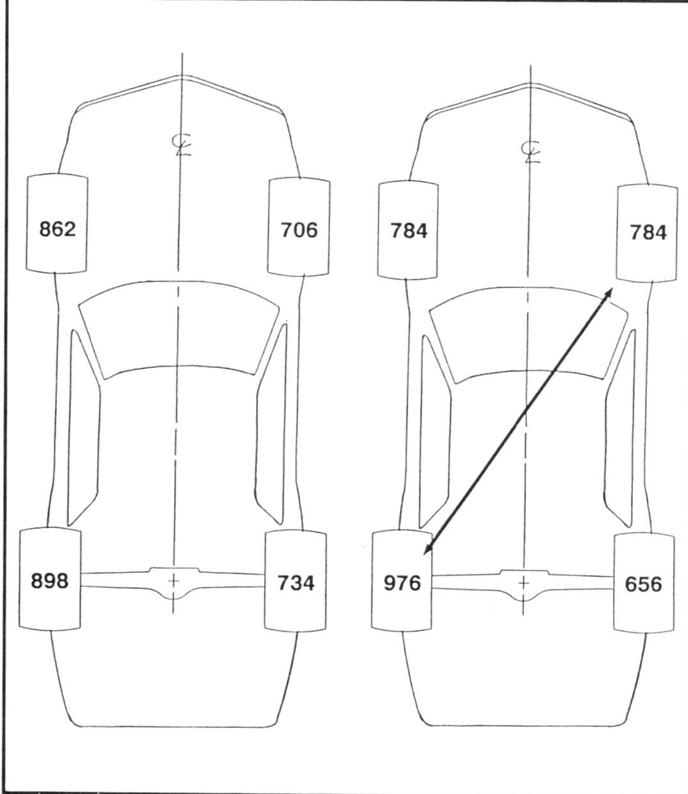

That is why we set the initial ride height 0.5-inch lower at these two corners.

To figure the desired corner weights which will yield the target cross weight: 1) Multiply the total weight of the car (3,200) by the desired cross weight percentage (55% or .55), which is 1,760. 2) Subtract from 1,760 the total of the cross weight we now have (706 plus 898 = 1,604), so 1,760 - 1,604 equals 156. 3) Divide 156 by 2, which is 78. 4) 78 is the number of pounds we want to add to the right front and left rear corners by screwing down equally on the weight jackers at those corners. If you have to change the corner height at the left rear and right front by more than 0.5-inch, screw UP equally on the left front and right rear weight jackers until the proper cross weight is achieved. After doing this, you will find that this added weight at the right front will have been subtracted from the left front, and the added weight at the left rear will have been subtracted from the right rear (assuming that your chassis is absolutely rigid).

The target corner weights for this car, with 55 percent left weight, 51 percent rear weight and 55 percent cross weight will be:

784	784
976	656

Notice here that the total of the left side corner weights is still the same (1,760) as it was in the previous example

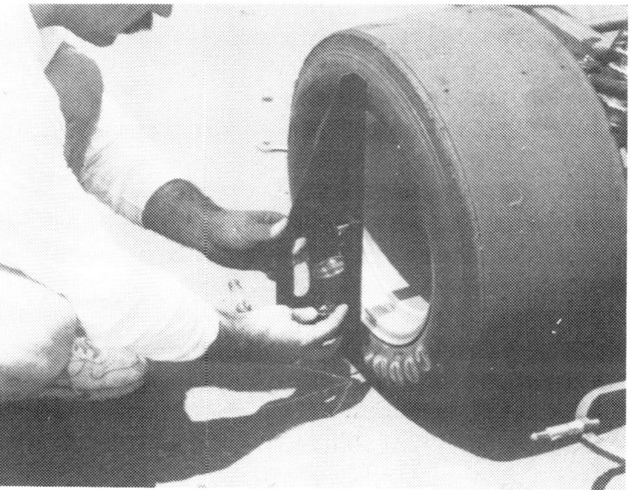

This type of caster/camber gauge, from Mitchell Engineering, is much easier to use in the field than the spindle-mounted type.

when we only had 50.1 percent cross weight. And the total of the rear corner weights is still the same even though the left rear weight has been increased substantially with the cross weight adjustment.

This same weight set-up procedure can be used on a car you have been racing for awhile if you want to start over and get a new chassis set-up baseline. Just follow the same procedure as we did above. Take all the ballast out of the car, and adjust the ride heights at all four corners to the desired numbers. Then put the car on wheels scales and distribute the ballast and set the cross weight as discussed above.

Front End Alignment

Caster

Setting the caster should be done before the camber because camber is affected by caster.

Caster provides directional steering stability. This influence is created with a line which is projected from the steering pivot axis down to the ground. This line strikes the ground in front of the tire contact patch when the caster is set in the positive position. A torque arm then exists between the projected steering axis pivot line and the center of the tire contact patch. The torque arm serves to force the wheel in a straight ahead direction. The greater the length of this torque arm (caused by greater amounts of positive caster), the greater the steering effort required to turn the wheels away from their straight ahead direction.

The difference in caster setting between the left front and right front is called caster stagger. A slight amount of caster stagger helps the car change from its straight ahead path more easily to ease into a left turn (when the caster at the right front is greater than the caster at the left front).

The caster stagger on a dirt track car is generally not very large because the same factor which aids in turning the car to the left increases steering effort when turning to the right to countersteer.

The amount of caster and caster stagger used on a race car is also influenced by the use of power steering. Cars with power steering can use more positive caster and more caster stagger. Power steering also allows dirt track cars to use more caster stagger.

Some racers utilize negative caster at the left front in order to gain more caster stagger. The car will track better and have better forward directional control with negative caster at the left front, but it is very difficult to drive a car with negative left front caster without having power steering.

You should be aware that the Camaro chassis with the angled swing axis of the A-arms will develop a little more than .5 degrees of caster gain per inch of travel.

For our blueprint chassis car, use the following guidelines for the initial caster settings:

Paved Track Car, Manual Steering
Left front: –.5° Right front: +1.5°
Paved Track Car, Power Steering
Left front: +1° Right front: +4°
Dirt Track Car, Manual Steering
Left front: +.5° Right front: +3°
Dirt Track Car, Power Steering
Left front: +1° Right front: +4°

Camber

The idea of the camber adjustment is to keep the tire contact patch flat on the track surface at the maximum point of cornering. Camber has the biggest influence over the vehicle's cornering ability than any other alignment feature. On most Pro Stock cars, the static camber setting at the right front tire is between 2 and 3.5 degrees negative, depending on the camber change curve of the suspension, the type of track and banking, and the tire construction.

For our blueprint chassis car, use the following guidelines for the initial camber settings (track testing may show that initial settings may need to be changed):

Paved Track Car
Left front: +1.75° Right front: –3.5°
Dirt Track Car
Left front: +1.75° Right front: –3.5°

Tire temperatures taken after practice laps will help you determine the exact camber requirements for your application. See the Tire chapter for more information on reading tire temperatures.

Toe-out

The amount of total toe-out that is correct for your application depends on the car's front track width, amount of Ackerman steer and the turn radius of the race track. The basic guideline for toe-out is 1/8 to 1/4-inch out. The tighter the track and turn radius, the more toe-out required. For our blueprint chassis car, we reccommend an initial setting of 1/4-inch out, for both dirt and pavement cars. Tire temperatures taken after some practice laps will help you determine the exact toe-out requirements for your application.

Some Hints For Working With Springs

Spring rates change, both in rate and height, after use. Variations are caused by the quality of the spring material and the type of use the spring receives. If you are unaware of the spring rate and height changes, it can adversely affect the handling of your car, and leave you wondering what went wrong.

When you first install new springs, use a quality spring rate checker to double check the rating (this is a good procedure to follow to guarantee that you got the spring rate you ordered), and measure the free height of each spring. Write these down in your chassis notebook, so you have a baseline to compare against in the future. Then periodically take the springs out and rate and measure them to see if there have been any changes. We have rated the springs on many race cars we have worked on, and have found that the actual spring rate is as much as 25 percent softer than the racer thought he had!

Starting Specs

The Recommended Starting Specifications charts show basic "ballpark" starting set-ups for an "average" dirt track and an "average" asphalt track. These numbers are based on a 3/8-mile oval, with up to 10 degrees cornering banking. This set-up can work on a 1/4-mile and a 1/2-mile track as well, with probably a few adjustments. These set-ups are also a little on the "soft side". A softer set-up requires a smoother driving style and a more experienced driver. A lesser experienced driver might want a slightly stiffer chassis set-up until he gets more track time and learns the feel of different handling conditions. And, the inexperienced driver would do best with a car that has a little more push in the chassis — it won't be the fastest way through the turns, but it will be a stable feeling car.

Recommended Starting Specifications: Dirt Track

Track Type:	Flat to med. bank 3/8-mile
Car Weight:	3,200 pounds
Weight Distribution:	53% left, 53% rear, 49% cross
Spring Rates:	
LF-975	RF-975
LR-200	RR-200
Shock Absorbers:	
(PRO Shock first 2 numbers)	
LF-76	RF-77
LR-95	RR-95
Front End Alignment:	
Caster (manual steering)	
LF +.5°	RF +3°
Caster (power steering)	
LF +1°	RF +4°
Camber	
LF +1.75°	RF –3.5°
Toe-out: 0.25-inch	
Ride Heights:	
LF-4.5"	RF-5"
LR-5.5"	RR-6"

Recommended Starting Specifications: Paved Track

Track Type:	Flat to med. banked 3/8-mile
Car Weight:	3,200 pounds
Weight Distribution:	55% left, 51% rear, 55% cross
Spring Rates:	
LF-975	RF-1100
LR-250	RR-225
Shock Absorbers:	
(PRO Shock first 2 numbers)	
LF-78	RF-78
LR-96	RR-96
Front End Alignment:	
Caster (manual steering)	
LF –.5°	RF +1.5°
Caster (power steering)	
LF +1°	RF +4°
Camber	
LF +1.75°	RF –3.5°
Toe-out: 0.25-inch	
Ride Heights:	
LF-4"	RF-4.5"
LR-5"	RR-5.5"

Your car develops a push going into the turns. What do you do? Follow the step-by-step guide below.

Chassis Adjustment At The Track

Sorting out chassis problems must be done in an orderly manner. Because turn entry problems can affect the car's handling all the way through a turn, and off of it, solving problems must begin here. Start at the braking point into the corner, then turn-in entry, corner apex, and corner exit. The car's handling should be thoroughly examined at each of these phases, and in that exact order. When solving problems, fix them as they appear in that order. Many times fixing a problem at the turn entry will cure a problem at the apex and turn exit.

Let's say you have a balanced car with a known chassis baseline after setting up the car in your shop. At your regular track you develop a handling problem of a push going into the turns. What do you do, and in what order, to correct the problem? Try these adjustments in this order (and be sure to make just one change in the chassis at a time):

1) Use more rear stagger
2) Take preload out of the front anti-roll bar
3) Take some cross weight out of the chassis. When doing this, adjust each corner of the car to keep the ride heights balanced. Screw down a little on the left front and right right, and screw up a little on the right front and left rear.
4) If the problem is more severe, and the above adjustments do not cure the problem, then a spring rate change should be made. Decrease the right front spring rate to the next lowest spring rate (usually work in 50 #/" increments at the front corners and 25#/" increments at the rear corners). Now test the car again.
5) If the problem still exists, there are other areas to look at. Check the tire temperatures, and see if there is an indication of a front end alignment problem, such as not enough negative camber at the right front. Also check the average temperature of the left front tire. Is it too cold? If so, move some ballast up the left side frame rail to put more

physical mass at the left front corner. Before you move the ballast, be sure to measure the corner heights of the chassis so the same balance can be restored after the ballast move.

6) Another area to look at is rear roll understeer. This is caused by the rear suspension linkages moving the right rear tire ahead and the left rear back during body roll. A good indication for the driver of this type of problem is that the understeer increases as the body roll increases. The understeer would be proportional to the amount of body roll because it is the body roll that is inducing it. The cure is to reset the linkage lengths or front attachment heights so that the rear end is not steered by linkage movement.

7) Does the understeer occur under braking during turn entry? The problem could be too much front brake bias. Adjust the brake proportioning bar to move more braking bias to the rear.

If the car is loose instead of pushing, follow this same sorting procedure as outlined above, but just make the adjustments opposite of what was suggested to cure the push condition. For example, first try decreasing rear stagger, then adding some preloading to the front anti-roll bar, then add cross weight to the chassis. If you need to change spring rates, your first change should be to increase the right front rate.

Using The Rear Roll Center As A Fine Tuning Adjustment

Adjusting the roll centers adjusts weight transfer from left to right. It is very difficult to change the front roll center, but on a car that uses a Panhard bar or Watt's linkage in the rear, it is very easy to move the rear roll center up or down.

Lowering the rear roll center increases side bite at the right rear (on dirt) or increases down force and traction on the right rear (on a paved track). The rear roll center can be used to adjust the amount of side bite. If the rear roll center is too high, and the front springs are too stiff, the car will go into a four-wheel drift. (The same is true if both the front and rear roll centers are too high.) If you lower the rear roll center, you will achieve more side bite, and probably cause a push in the chassis. Then you can go to a softer spring rate in the right front to balance the car. The overall effect is a little more body roll can more side bite.

The rear roll center can also be used to adjust a push out of the car. Move the Panhard bar up, which raises the rear roll center, and this takes away from bite at the right rear which will take car of the push. Or, if your right rear spring rate is too stiff (causing a loose condition), and you don't have the proper spring rate available, you can lower the rear roll center a little to improve the handling.

This same technique can be used to adjust a dirt track car that races on a track which starts out real wet and ends up real dry. For the wet track condition, start with the Panhard bar higher, then move it downward as the track dries out.

Trouble Shooting Chart

Car Is Loose Going Into A Turn

Item	Adjustment
Stagger	Right rear is too large
Toe-out	Not enough
Front springs	Too weak
Rear Springs	Too stiff
Anti-roll bar	Too weak
Brake bias	Too much rear
Suspension bind	Exists in rear suspension
Cross weight	Increase

Car Pushes Off Corner

Stagger	Increase right rear size
Toe-out	Decrease
Front springs	Too stiff
Rear springs	Too weak
Anti-roll bar	Decrease rate
Cross weight	Use less

Note: This is a very basic quick reference guide to common handling problems. The charts are not meant to cover all problems and situations, but rather to give you an idea of where to start when you experience a problem.

Banked Track Corrections

Going to a medium-to-higher banked track from a flat-to-slightly banked track requires several changes in the chassis set-up. First, banking creates downforce, so all four spring rates have to be stiffer to resist it and hold the car up.

The banking and downforce will also increase rear tire bite. To overcome this, the rear spring rates have to be increased, especially at the right rear. If you have a problem getting the car to turn, increase the right rear rate more. You can fine-tune the chassis by adding more stagger (but don't use stagger as a crutch to cure other problems, such as spring rates).

Be careful of getting the overall chassis set-up on a banked track car too soft. A soft set-up has more body roll, and banked tracks are faster and make things happen quicker. The result is that a soft set-up can end up reacting too quickly.

For a higher banked track, you want a lower roll center and less negative camber gain, because there is much less body roll and there is more downforce. On a flatter track you have much more body roll, so you need more negative camber gain which you get with the higher roll center.

Suppose you have a car that you race at two different tracks — one a paved 3/8-mile track with 5 degrees banking, and the other a paved 1/2-mile track with 17 degrees banking. The lower banked track would require a 4-inch roll center for more negative camber gain at the right front. The higher banked track would require a 2.5-inch roll center to decrease the right front negative camber gain.

Both of these can be accomplished by having an upper A-arm mounting plate with two sets of mounting holes, so the A-arm inner pivot point can be moved up or down to affect the roll center change. The upper mounting position would be for the 2.5-inch roll center, and less negative camber gain; the lower mounting position would be for the 4-inch roll center and more negative camber gain.

Using Shock Absorbers To Fine-Tune The Chassis On Dirt

Shock absorbers will help speed up or slow down the rate of weight transfer that occurs at any corner of a car. Shocks will not increase or prevent weight transfer, but rather they alter the rate of speed at which that transfer occurs. They can help hold up or hold down a corner of the car for a longer period of time while weight transfer occurs.

On a very rough short track, it is better to soften the rear spring rates slightly, rather than to soften up the shocks. This lets the rear wheels of the car follow the undulations of the track, yet the shocks will provide enough control, especially under braking entering a corner. It is better to have a car over-shocked than under-shocked.

For a dry slick dirt track, shocks can be changed to help encourage faster weight transfer from the front wheels to the rear for a better rear side bite and forward bite. At the left and right front, use 75 shocks. This will help unload left rear weight onto the right front on corner entry to get side bite to allow the car to drive in low. This will also expedite weight transfer from the right front to the left rear under acceleration for a better bite coming off the corner.

Under extreme conditions, try a right front shock which has a different damping rating on compression than rebound. What is needed is a shock that has sufficient damping force to hold the car up during corner entry, but pops up easily under acceleration and transfers weight quickly to the left rear. Such a shock would be the Afco 1075-3 or Carrera 3173/5 which have a 5 rating on compression and a 3 on rebound. If more transfer and left rear loading is required, put a 1075-3 or 3173/5 shock on the left front as well.

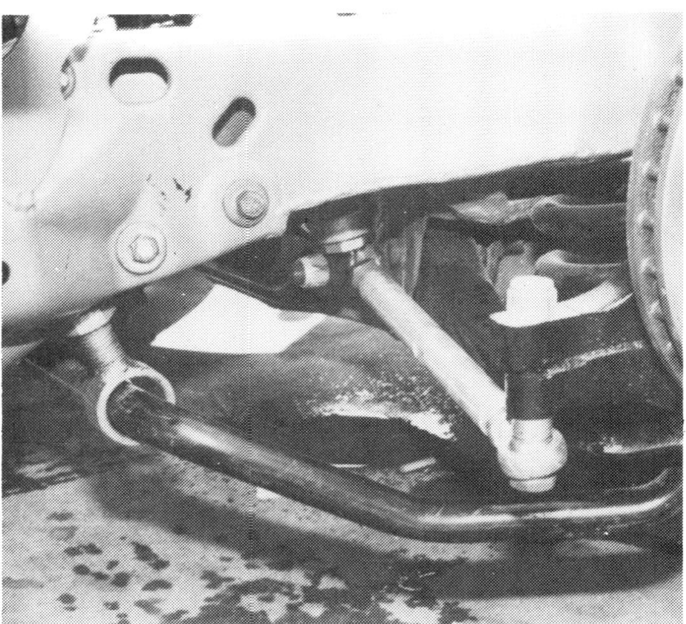

This type of anti-roll bar attachment will allow you to preload the bar to fine tune the roll couple.

Roll Couple Distribution

Roll couple is the force, due to cornering, acting on the sprung weight of the car, rolling it about the roll axis. The front-to-rear handling characteristics (understeer or oversteer) of a race car can be tailored or adjusted by adjusting the front-to-rear roll stiffness proportioning. This is called roll couple distribution. The end of the car with the highest roll stiffness will receive the greatest amount of weight transfer during body roll. The higher the front roll couple percentage, the stiffer the front roll stiffness is, which means more of a tendency toward understeer. Conversely, the less front roll couple, the more oversteer.

The best way to fine tune the roll couple distribution is with an anti-roll bar rate change. A racer should have at least three different rates of anti-roll bars available to him to make front stiffness changes. By changing only the front anti-roll bar rate to change oversteer or understeer, we are not making any other change which effects any other facet of the chassis. Adding pre-load to the anti-roll bar also acts as a fine tuning adjustment to add more rate at the front, or taking out pre-load decreases front rate.

Chapter 10

Safety

Safety is a difficult item to sell to racers. If someone could come up with a magic helmet or driving suit that would guarantee a driver another 3 MPH, racers would beat a path to his door.

But safety doesn't make the car go faster. In fact, it's a boring subject to racers. But it is something every racer MUST pay attention to. His survival depends on it. Don't skip this chapter.

Whenever you are building any type of race car, safety should always be the primary concern. Anyone who has ever attended a motorsports event is acquainted with the dangers of auto racing. It would appear that the racer would be the most responsive to the needs of safety equipment, designs and procedures in his race car — but apparently not so.

While most racing organizations have some rules relating to safety, clear rule book definitions and tech inspections are often lax or even nonexistant. Terms such as "approved" still appear in rule books when relating to fuel cells, driving suits or safety harnesses. Other orgainizations like SCCA and IMSA have stricter rules and in-depth safety inspections. The ultimate responsibility lies with the driver.

Safety equipment is designed and manufactured to prevent injuries to the driver in the event of an accident. But safety goes much further than that. Safety, driver protection and crash worthiness are all very much a part of how you design and construct your race car.

Construction Techniques

Proper construction techniques means preventing mechanical failures. This means not only thorough preparation, but also utilizing good welding methods and techniques, utilizing the proper size materials, and the correct grade of metals. Joining, fastening, and even painting play an important role in keeping a race car together. Using the proper fasteners in the chassis, such as the RBW L9 series of fasteners from ARP, adds greater strength and reliability. The L9 series of fasteners is still ductile but of greater strength than grade 8 fasteners.

Taking care that no parts of the suspension or steering bind throughout the complete range of the component's travel will also prevent problems. Good wiring techniques are also important. A loose wire can stall an engine at a critical moment, or worse yet, start a fuel or electrical fire. Very dirty or rusted parts can mask cracks while the rust can weaken the metal. Improperly designed or machined parts will not fit properly and could fail at a very inopportune time.

Good quality fasteners, and the correct size for the job, prevent part failures. Note the safety wiring on the shock mount for extra protection.

Using the wrong grade, type or size of fastener will lead to trouble. Magnafluxing or other forms of nondestructive testing should be performed on all critical parts of the steering and suspension. Firewalls should be as their name implies — fire proof. And don't leave holes in the firewall for fire to leak through.

The driver himself is often the most dangerous element on the track — both to himself and to others. Burning the midnight oil and expecting to drive at your best the next day is not very sensible. When tired, the mind reacts more slowly and judgment is impaired.

This same type of warning should go for the angry or revengeful driver. Short track racing at times can be a contact sport, and some guys like the contact more than the sport. Being the "contacted" driver can be infuriating, but

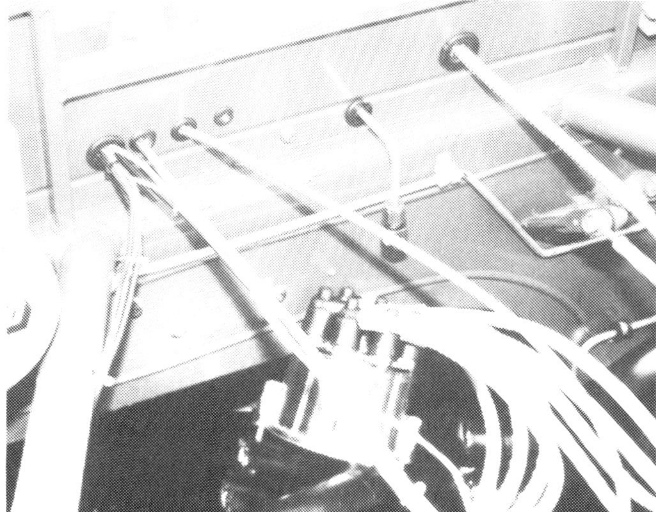
Notice how this well-prepared car has grommets around all tubes and wires that pass through the firewall. This effectively seals the driving compartment from the engine compartment, and prevents chafing of the wires and tubes.

driving to get even with another driver can only distract and lead to trouble.

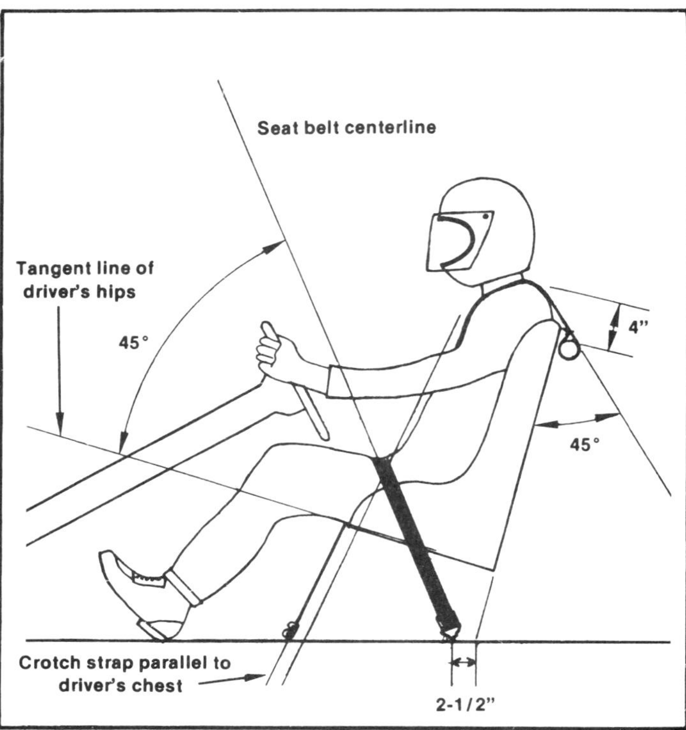

The correct mounting locations and angles for the driver restraint system.

Driver Restraint

Restraining the driver's body will prevent injuries from contacting the interior of the car, a wall, another car or its parts, or even the track surface. The restraint system consists of the lap belts, shoulder harnesses, the crotch belt, neck brace, helmet restraint strap, arm restraint and the window net. The restraint system also helps prevent injuries from the inertia generated by a flip or crash.

Are you thinking here that you race at just a short track where the speeds aren't too high, and the safety equipment we are specifying really doesn't have to apply to you? You are wrong. Take this seriously. Accidents can happen even at the slowest of speeds. If you follow the proper safety procedures and use the proper equipment (and always buy new, not used, equipment) your chances of serious injury are very much reduced.

Heat/Fire Restraint

The driver's firesuit is one of the most important safety item he has. Fireproof underwear, gloves and shoes are also important to use to prevent painful burns, but in the event of a fire the driver will want to escape promptly. The escape would be very difficult if the driver's hands and/or feet are burnt.

A good firewall and sheet metal barrier will help prevent fire from entering the driver's compartment, and a good fuel cell (which conforms to NASCAR's stringent rules) will take quite a shock before bursting. On-board fire suppression systems can help put out a fire in a hurry, or at least allow the driver plenty of time to abandon the car.

Preventing a fire is still the best way to fight a fire. Don't make the mistake many builders make by using cheap fuel lines and hose clamps while following the rules by using a good fuel cell. Also, purchase your fuel cell from a reputable manufacturer, and install it according to their recommendations.

Driver Impact Prevention

The best way to prevent the driver from making contact with the interior of the car is to take the driver's height and size into consideration when constructing the car. Allow a minimum of four inches between the top of the roll bars and the top of the driver's helmet.

The rest of the roll cage should be located as far as possible from the driver's head and arms. Remember that under severe impact the neck and spine can stretch as much as ten percent, and a belt can stretch as much as two inches per foot. All that combined can cause the driver to reach out of the window or over to the passenger side. Or, worse yet, the driver can reach over the roll cage and contact the track surface if the car is upside down.

The roll cage structure should be located as far as possible from the driver's head. Note how the upper bars are padded with high density foam anywhere the driver's head can possibly reach. Also note the steel plate above the upper roll bar (see arrow). It gives additional protection to the top of the car should it get upside down.

Note the generous padding to protect the driver. Also note the excellent "X" bracing of the chassis inside the center bay of the car.

Contact with the interior of the car can be cushioned with a high density foam padding, such as the Longacre 6502, to improve shock absorption. This foam padding will also make entry and exit of the driver's compartment more comfortable. Be sure to put a section of the foam padding around the gear shift lever too.

A window net will help keep the driver's arms and head in the car should his body stretch that far, and the net will also prevent debris from hitting the driver. The window net attachment to the car should also be carefully considered. NASCAR has recently required the use of a passenger car seat belt attachment clamp sewn to the window net at the top as a quick release mechanism. Make sure the window net fits snugly when it is in place so that stretch in the material is tight enough to prevent the driver's head or arms from going outside the window opening plane during an impact.

This is the current NASCAR requirement for positive quick release mounting of the window net. The upper retaining rod for the net is held at the rear in a spring loaded bracket, and at the front by a seat belt latch which can easily be reached by the driver from inside the car.

The window net should be very easy for the driver to open and remove from inside the car all by himself, even if there has been some damage to the driver's side of the car. Be sure your driver can do it. And, make sure that all clamps and brackets which hold the window net in place will sustain a heavy blow and still keep the net in place.

Interior Sheet Metal

Make sure that the interior sheet metal is installed in such a way that the driver's head cannot contact it. This sounds simple, but keep in mind that under a hard impact the human body is very elastic. The best idea here is to construct the race car interior so that the interior sheet metal falls at least eight inches below the shoulder of the driver.

There are two reasons for proper sheet metal placement. First, if the sheet metal is even shoulder high, there is the possibility of the driver's head contacting it in a hard crash. In a very hard crash this can cause severe injury — or worse. Second, high sheet metal placement can block a driver's exit out the right side of the car. After a wreck, this might be the only exit path available to the driver. Would you want this path blocked if the car is on fire?

The method of sheet metal attachment to the chassis is also very important. Sheet metal pieces and panels have to

The arrow indicates an electrical kill switch in this car. Note the clean layout of the cockpit.

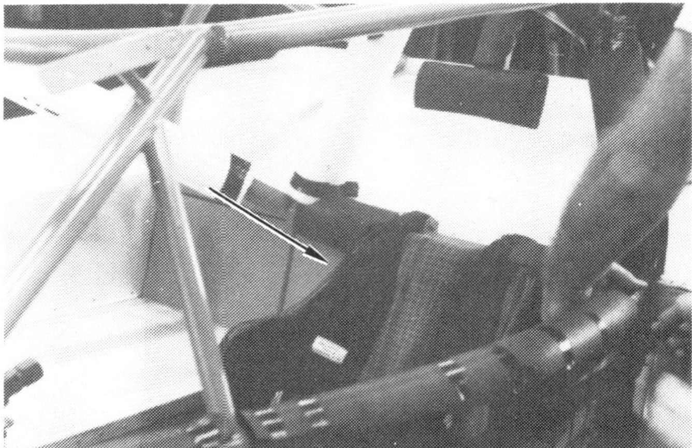

This racing seat has right side lateral bracing all the way up to the shoulders to help the body handle side loads AND side impacts.

be prevented from popping loose and flying around in the cockpit during a wreck. Pop rivets can't be relied on under impact. They will fail. It is far better to use quality quick release fasteners. The sharp edges of loose sheet metal panels can become lethal weapons during a wreck.

Other Cockpit Items

The ignition components, fire extinguisher, radio and other attachable items must be located where the driver cannot possibly contact them under a hard impact, or where the driver can get caught up on them during a quick exit out of the car.

Quick release steering wheels are a great idea. But, just in case the quick release mechanism is jammed by impact from a wreck, make sure the driver can get out from under the steering wheel if it cannot be released.

It is a good idea to use a master electrical kill switch which shuts off all electrical systems, even if it is not required by your association rules. These are required by SCCA and IMSA. The cut-off switch must be installed within the reach of the driver to faciliate complete shut-down of the electrical system in case of an emergency. The shut-off switch should be clearly marked to be visible from the outside of the car, so it is readily accessible to rescue workers in case of an emergency. The shut-off switch also comes in handy when working on the car's electrical system and to shut the whole system off when the car is in storage. A switch of this type is available from Moroso, part number 74100.

Escaping a crashed race car can be time consuming, especially if a driver is stunned or injured. The safety minded driver should practice a logical sequence of escape from his car, and store the information in the back of his mind. Frequent mental imagery of this procedure will program the driver's mind to react to an emergency, so even under the most extreme emergency the driver will be able to react without even thinking about what needs to be done.

The driver has the responsibility every time he enters his race car to survey every inch of the cockpit. Even though he is familiar with it, he should look for anything that has been changed or loosened, tools that have been left inside, panels that are loose, etc. The driver cannot ever take anything for granted or become complacent.

Seats and Mounts

The racing seat should be mounted to the chassis via a steel tubing seat hoop. The seat hoop should be constructed of 1.25-inch O.D., 0.125-inch wall round steel tubing, and should be mounted to the left on the chassis frame and to the roll cage behind the seat. The seat hoop should also be welded in a third location near the right side to prevent pulling up in case of an accident. Such a design will cause the seat to move with the chassis in the event of a hard impact to the left side of the car. Any plates or brackets welded to the seat hoop should be a minimum of 3/16-inch thick, and care should be exercised not to drill mounting holes too close to the edge of the bracket to prevent fatigue cracking from the hole.

With a fiberglass seat, use 3/8-inch O.D., 0.065-inch wall round tubing to form a supporting frame around the complete perimeter of the seat, across the back and down the center of it. This frame gives secondary support for the driver should an impact ever break the fiberglass seat

The seat must be mounted at six positions on the seat hoop, four on the bottom of the seat and two at the upper back of the seat. Be sure to use rounded head bolts to insure driver comfort when mounting the seat, and use large

The seat attaching hoop is mounted to the frame on the left and the roll cage behind it.

This 3-layer driving suit is cool even in hot weather and it provides the driver excellent protection. A good helmet, gloves, fire retardant shoes and underwear are good insurance. The backing on these gloves is of Nomex material and is dyed red so hand signals can be seen clearly. A visor or goggles should be worn even in a closed car to prevent eye damage.

washers under each bolt head to prevent the bolts from pulling through the seat, or work-fatiguing the seat.

The seat should provide an even amount of support to the thighs, hips, torso and shoulders. The oval track seats used in stock cars for so many years are not as safe as seats used in road racing cars. The typical stock car seat provides support under the arm pits against the ribs. In a heavy side load impact, the ribs will be heavily bruised, or worse, broken.

The typical road racing seat spreads the load over the lower part of the body, torso and the shoulders. The human shoulder is much better equipped to take side loads than is the rib area.

Remember that the basic function of the seat is to maintain the driver in a certain predetermined position during a race that is comfortable to the driver. And, in case of an accident, it must be strong enough to support the driver and keep him restrained in position. Should the seat deform or break up, the driver will no longer be held in place by the safety belts and harnesses.

A seat, simply bolted to the hoop, will tear right out of the mounting bolts under a severe impact. The steel tube frame should provide support to the seat and relieve the stress from the mounting surfaces.

The Headrest

A good headrest, in conjunction with a neck brace, will help prevent injuries to the neck and head in case of an accident. A good headrest should be curved to conform to the curvature of the neck, especially at the base of the skull. In a rear end collision, or if the car should hit the wall backing up, a perfectly flat headrest will come in contact with the rear of the helmet first, pushing the head forward and the neck back. This will cause whiplash and possible serious injury. A well designed headrest should provide support to the neck and the rear of the head to prevent the whiplash effect.

The headrest should be sturdy but giveable, and it should never be constructed with pieces of tubing pointing toward the rear of the driver's head (what if the headrest breaks off the tubing during a flip, and all the driver's head has to contact on the next impact is the open end of the tubing?).

Driver Protective Clothing

NEVER drive a race car without some form of a fire and heat retardant driving suit. There are several different types and qualities of material used for driving suits, and the driver should research the various types and make himself familiar with all of them before making a purchase. The different types and layers of fire retardant fabric will provide varying amounts of thermal protection to the driver.

Fire retardant underwear (socks, Long Johns, long sleeve tops and face hoods) will add several seconds of safety if the driver is exposed to fire. Areas of the body such as the neck, ankles and wrists are often burnt when the driver does not wear fire retardant undergarments. Full underwear is highly recommended to drivers using only a single layer driving suit. However, a multi-layer suit and under-

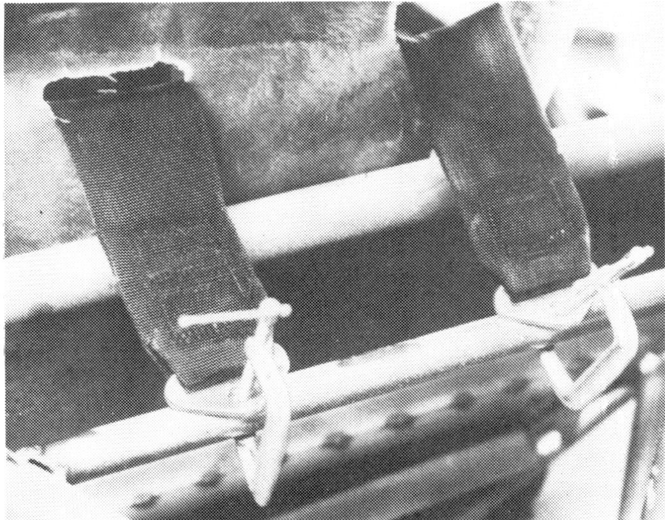

The shoulder harnesses should be attached close to the driver's shoulders to keep the lengths as short as possible. Make sure that when harnesses go through a fiberglass seat the openings in the seat are protected with seat covering material. The rough edges of the fiberglass will chew through the belt material.

wear will greatly enhance the chance of escaping injury from fire and heat.

Gloves are important in a fire to allow the driver to remove his equipment before bailing out without burning his hands. If the driver burns his hands, he will not be able to release his harness or window net, or even get out of the car.

The same goes for the shoes — the driver will have a hard time escaping a burning car on blistered or burnt feet.

It is highly recommended that the driver take advantage of all the protective clothing that is available. It can save his skin.

Belts and Harnesses

Belts and harnesses must absorb a tremendous amount of load in accidents. To be sure of getting the best for your money only purchase belts from reputable U.S. manufacturers, such as Pyrotect, Simpson or Deist. Some foreign harnesses are plenty strong, however the buckles cut the fabric will before the maximum allowable forces are encountered.

One of the most common mistakes with harnesses on stock cars is that the installers use much too long of a belt or harness. A 3-inch belt can stretch as much as two inches per foot in a good size crash. Head tilt, neck stretch, slack in the harnesses and two inches per foot of belt stretch can easily allow a driver to hit the steering wheel, or worse, the driver can slip out of the harnesses altogether.

Another thing to avoid is installing the shoulder harness mounts too far apart from each other. This will cause the harnesses to slip off the driver's shoulders during a race. The double strap type of shoulder harness (each strap having its own mount instead of being sewn together in a "Y") is much stronger and should be used. The double harnesses should mount behind the driver from four to six inches apart with a short length of strap behind the driver to prevent unnecessary strap stretch.

Neck Braces

A neck brace will not only absorb some of the movement generated by the head in an accident, but it will also relieve some of the strain on the neck stemming from "G" loads in the corners. The full face helmets are notorious for breaking collar bones. What happens is that when the car hits something, the car slows down but the driver keeps moving.

The body is help solidly in place by the seat and the restraint system, but the head, with the added weight of the helmet, will pivot about the neck and the lower lip of the helmet contacts the collar bone with enough force to break it. A good neck brace will cushion the shock to help prevent injury.

Helmets

The helmet is probably your most important piece of safety equipment. Don't take any shortcuts here. Buy only the best. If your budget won't allow it, don't race until you can afford to buy an excellent one. Be sure to purchase a helmet that bears the latest "Snell" sticker and you will comply with all the latest safety requirements for helmets.

Follow these tips with your helmets:

1) When purchasing a helmet, make sure it fits snugly, but not too tight, as headaches may result. Keep in mind that your head will swell slightly on a hot day. The helmet should not be so loose with the chin strap off that it can rotate on your head while you try to twist it.

2) When adjusting your helmet before going out on the track, always tighten the strap as tight as you can stand it and press down on the top of the helmet to settle it on your head. Just like with seat belts, the chin strap will loosen up in a few minutes from the helmet settling.

3) Use goggles or a shield to protect your eyes from dust and flying debris, but use caution when using a full face helmet and a shield. In an enclosed car, the driver may not get enough ventilation to replenish the oxygen supply inside the helmet. This situation is worsened by the use of a neck brace or head sock.

4) Do not paint, sand or drill on a helmet. Some paints will attack the fiberglass, while sanding and drilling will weaken the structure.

Suppliers Directory

Note: Telephone area codes are changing throughout the country. If you call a number for a company shown here and the number is not active, check if the area code has changed.

AFCO
(American Fabricating Co.)
P. O. Box 548
Boonville, IN 47601
(812) 897-0900
Coil springs, torque arms, brake calipers and rotors, spherical rod end bearings, torque arms, shock absorbers, birdcages, trailing arms

Borgeson Universal Company, Inc.
187 Commercial Blvd.
Torrington, CT 06790
(860) 482-8283
Universal joints for all applications in racing

CSC Racing Products Inc.
125-A Harry Walker Pkwy
Newmarket, ON L3Y 7B7
Canada
(905) 954-0522
Roll cage kits, interior sheet metal kits, coil springs

Carrera Shocks
5412 New Peachtree Rd.
Atlanta, GA 30341
(770) 451-8811
Shock absorbers

Coleman Machine Inc.
N-1597 US 41
Menomonie, MI 49858
(906) 863-8945
(800) 221-1851
Suspension components

Dan Press Industries (DPI)
1397 St. Hwy 506
Vader, WA 98593
(360) 295-3436
Gold Trak differential

Ebeling Engineering
11701 McBean Dr.
El Monte, CA 91732
(626) 442-0953
Suspension components and race car chassis fabrication

Fuel Safe Racing Cells
63257 Nels Anderson Rd.
Bend, OR 97701
(541) 388-0203
Racing fuel cells

Hedgecock Racing Entrerprises
1520 Horneytown Rd.
High Point, NC 27265
(910) 887-4221
Race car chassis fabrication, components

Howe Racing Enterprises
3195 Lyle Rd.
Beaverton, MI 48612
(517) 435-7080
Spindles, hubs, brakes, chassis components

Hyperco
810 Bates Rd.
Logansport, IN 46947
(219) 753-6622
Racing coil springs

Irvan-Smith, Inc.
1027 Central Dr.
Concord, NC 28027
(704) 788-2554
Bump steer checking gauge, metal fabricating equipment

JOES Racing Products
11400 Airport Rd.
Everett, WA 98204
(425-267-9199
Caster/camber gauge and turn plates, clamp-on brackets, stagger caliper

Lee Manufacturing
11661 Pendleton St.
Sun Valley, CA 91352
(818) 768-0371
Power steering kits

Lefthander Chassis
13750 Metric Dr.
Roscoe, IL 61073
(815) 389-9999
Race car chassis, components

Longacre Automotive
14269 NE 200th St.
Woodinville, WA 98072
(425) 485-0620
Roll bar padding, tire pyrometers, tire durometers, coil spring checkers, computerized chassis scales, dash panel kits, tire pressure relief valves, electrical switches

Marsh's Racing Tires MRT)
22624 Marsh Rd.
Siloam Springs, AR 72761
(501) 524-4157
Racing tires, tire sealant, tubes, wheels

Moroso Performance Products
80 Carter Drive
Guilford, CT 06437
(203) 453-6571
A wide variety of parts necessary for chassis and engine building

PRE
(Professional Racers Emporium)
2376 Main St.
Riverside, CA 92501
(909) 779-1300
Complete race car chassis and fabrication, frames and cages, suspension linakges, steering linkage, spindles, suspension parts

PRO Shocks
1715 Lakes Parkway
Lawrenceville, GA 30043
(770) 995-6300
Shock absorbers, coil-over units, coil-over springs

QuickCar Racing Products
44 Pearl Pentecost Rd.
Winder, GA 30680
(800) 997-7333
Gauge panels, switch panels, pit equipment, bump steer gauges, tire pyrometers, accessories

RPM Co.
957 W. 'A' St.
Hayward, CA 94541
(510) 783-3714
Race car mufflers, chassis components, chassis

Sierra Racing Products
1558 Forrest Way
Carson City, NV 89706
(775) 882-3500
Brake calipers, rotors, mounting brackets, pads, fluid

Smoky Joe Racing Chassis
800 Calhoun Ave.
Rome, GA 30161
(309) 232-5221
(800) 338-2197
Race car chassis, components

Speedway Engineering
13040 Bradley Ave.
Sylmar, CA 91342
(818) 362-5865
Race car chassis components

Speedway Motors
300 Speedway Circle
Lincoln, NE 68502
(402) 474-4411
They sell almost every part necessary for building a Pro Stock car

Stock car Products Inc.
10 Thurman St.
Richmond, VA 23224
(800) 262-8099
Roll cage kits

Suspension Spring Specialists, Inc.
803 Water St.
Bremen, IN 46506
(219) 542-2870
Racing coil springs

Sway-A-Way
20755 Marilla St.
Chatsworth, CA 91311
(818) 700-9712
Anti-sway bars

Sweet Mfg. Inc.
3421 S. Burdick
Kalamazoo, MI 49001
(616) 344-2086
Power steering systems

Tanner Racing Products, Inc.
116 Clay St. NW, B-2
Auburn, WA 98002
(253) 939-1415
Electronic wheel scales, caster/camber gauges and turn plates, stagger caliper, data acquisition systems, tire pyrometers, tire pressure gauges, gauge and switch panels, clamp-on brackets

Tilton Engineering
Box 1787
Buellton, CA 93427
(805) 688-2353
Brake master cylinders, pedal assemblies, brake bias adjusters, brake remote reservoirs, hydraulic clutch systems, heavy duty starters

Wilwood Engineering
4700 Calle Bolero
Camarillo, CA 93102
(805) 388-1188
Calipers, rotors, hubs, brake pads, brake fluid

Tech Software, Videos, Books

Tech Information That Puts You At The Top Of The Field

The Complete Karting Guide
For every class from beginner to enduro. Covers: Buying a kart/equipment, Setting up a kart — tires, weight distrib., tire/plug readings, exhaust tuning, gearing, aerodynamics, engine care, competition tips. **#S140...$16.95**

4-Cycle Karting Technology
Covers: Engine bldg. & blueprinting, camshafts, carb jetting & fine tuning, fuel systems, ignition, gear ratios, aerodynamics, handling & adjustment, front end alignment, tire stagger, competition driving tips. **#S163...$16.95**

Kart Racing – Chassis Setup
How a kart chassis works & how to improve it. Covers: steering principles, weight distrib., tire theory & use, chassis setup & alignment, trackside chassis tuning, & more. For 2 & 4-cycles, asphalt & dirt. **#S245...$16.95**

Kart Driving Techniques
By Jim Hall II. A world-class driver and instructor teaches you the fundamentals of going fast in any type of kart. Includes: kart cornering dynamics, learning the proper line, learning braking finesse, how to master trail braking, racing & passing strategies, getting out of trouble, & much more. For all drivers from novice to expert. Includes a chapter on shifter karts. **#S274...$16.95**

Beginner's Package

1) Building A Street Stock Step-By-Step
A complete guide for entry level stock class, from buying a car and roll cage kit to mounting the cage, stripping the car, prep tips, and mounting linkages, etc. **#S144...$15.95**

2) The Stock Car Racing Chassis
A basic stock car handling book. Covers: Under and oversteer, The ideal chassis, Spring rates and principles, Wedge, Camber/caster/toe, Guide to bldg. a Chevelle/Monte Carlo chassis. **#S101...$10.95**

3) So You Want To Go Racing?
Everything you need to know to get started in racing. **#S142...$11.95**

Special Offer: Buy all 3 books above together and the special package price is only $33.50. Shipping charge is $10.

The Racer's Math Handbook
An easily understood guide. Explains: Use of basic formulas, transmission & rear end math, engine, chassis & handling formulas, etc. Includes computer program with every formula in the book & more. **#S193C..$24.95**

The NEW Racer's Tax Guide
How to LEGALLY subtract your racing costs from your income tax by running your operation as a business. Step-by-step, how to do everything correctly. An alternate form of funding your racing. **#S217..$17.95**

Advanced Race Car Suspension
The latest tech info about race car chassis design, set-up and development. Weight transfer, suspension and steering geometry, calculating spring/shock rates, vehicle balance, chassis rigidity, banked track corrections, skid pad testing. **#S105..$16.95** WorkBook for above book: **#WB5..$10.95**

Racer's Guide To Fabricating Shop Eqpment
An engine stand, hydraulic press, engine hoist, sheet metal brake and motorized flame cutter all for under $400. Step-by-step instructions. Concise photos and drawings show you how to easily build them yourself. **#S145..$16.95**

Short Track Driving Techniques
By Butch Miller, ASA champ. Includes Basic competition driving, Tips for driving traffic, Developing a smooth & consistent style, Defensive driving tactics, and more. For new and experienced drivers alike. **#S165...$16.95**

Bldg. The Pro Stock/Late Model Sportsman
Uses 1970-81 Camaro front stub with fabricated perimeter frame. How to design ideal roll centers & camber change, fabricating the car. Flat & banked track setups, rear suspensions for dirt & asphalt, track tuning. **#S157...$21.95** Special package price for book & Bldg. The Stock Stub Race Car video is only $55.95

Sprint Car Chassis Set-Up
Detailed info on chassis set-up and alignment • Birdcage timing • Front end geometry • Choosing torsion bar rates • Chassis tuning with shocks • Using stagger and tracking • Tire selection, grooving, siping, reading • Blocking, tilt, rake, weight distrib. • Track tuning • Wings. **#S174 ...$19.95** Special Package Price for Sprint Car Chassis Setup Video & book is $53.95.

Mini Sprint/Micro Sprint Racing Technology
Covers frames, suspension, chassis adjustment & setup, blocking, wt. dist., chain drive systems, chassis squaring, tires, wheels, wings, & much more. **#S167...$17.95**

How To Build Chev Small Block V8 Race Engines
For the ultimate in affordable performance. Details every step of building: tear down, machine work, porting & cylinder head work, assembly, break-in, tuning & inspection. Affordable high perf. tips & techniques. **#S211...$19.95**

Hot Holleys
Extensive tuning & modification info on the Holley 4150HP & Dominator. Covers all CFM sizes of 4150HP. Shows complete gas & alcohol buildup of a 390 CFM late model/modified carb. Complete alcohol tech. **#S278..$19.95**

Trailers — How To Design & Build
All the tricks of the trade for building a race car trailer. Details on suspension, materials, construction, braking systems, couplers. **#S203...$24.95**

Trailer Building Blueprint #B800...$17.95 Special Package Price for blueprint & S203 book is only $36.95.

Carroll Smith's Engineer In Your Pocket
This is a handy, concise, indexed guide to analyzing and correcting race car handling problems. Organized by problem (such as corner entry understeer, etc.) with proven solutions. Almost every solution to every chassis problem is contained. 3.5"x7.5" **#S273...$15.95**

Chevy Small Block Part & Casting # Guide
Covers all small block heads, blocks, intake & exhaust manfolds from 1955 – 1990. **#S208..$12.00**

How To Run A Successful Race Team
Covers key topics such as team organization and structure, budgeting & finance, mngmt., team image, goal setting, scheduling & more. A valuable asset for all racers. **#S265...$17.95**

Think To Win
By Don Alexander. More than just a driving manual, it teaches the all-important mental aspects of racing – strategy & tactics, how to direct your focus, how to identify & correct mistakes – all the things winning drivers do instinctively. **#S231...$19.95**

COMPUTER PROGRAMS

Computerized Chassis Set-Up
This WILL set-up your race car! Selecting optimum spring rates, cross weight and weight dist., computing CGH, front roll centers, stagger, ideal ballast location, final gearing & more. Windows 95/98 Format. **#C155... $99.95**

Racing Chassis Analysis
You won't ever need a bump steer gauge again! This program does a complete bump steer analysis with graphs, camber & roll center change. Plus a complete dynamic cornering analysis. **#C176...$99.95**

Engine Shop Computer Program
For most all small & big block Chevy, Ford & MoPar V8s. Build your engine on the computer with cataloged parts. Then it simulates a dyno & plots torque & horsepower output. Then change a part & plot the difference. Very accurate! Windows 95/98 **#C216..$74.95**

TirePro Computer Program
Quickly & easily performs a detailed analysis of tire temps, providing info on: Tire pressure, camber, weight, springs & more. **#C178...$49.95**

Race Car Simulator – 2
Shows effects of chassis changes. Specify chassis & track specs, then the computer shows how the car will handle. Change springs or weight and the computer tells the new handling. Windows 95/98 **#C247...$99.95**

Front Suspension Geometry PRO
Does 3D analysis & graphs all suspension movements including camber curve, caster curve and bump steer. Makes suggestions on what parts to move & where to improve suspension geometry. For all cars, Windows 95/98, on CD-ROM. **#C249...$119.95**

Order Hotline: 714-639-7681
Order from our Web site: www.ssapubl.com

The Great Money Hunt
This book is ONLY for those who are really SERIOUS about obtaining sponsorship. Provides strategy with proposals, justifications & sample correspondence. How to find prospects, & sell and fulfill sponsor's marketing needs. Detailed & comprehensive. **#S200...$69.95**

Get Sponsored
By Charlie Hayes How to create a sponsorship program & proposal with a 7-step approach. A guide to finding prospects, sample correspondence, contract agreement. **#S260...$59.95**

Brand Fans – What Sponsors Really Want
Learn what sponsors really want from a motorsports sponsorship program. How to capitalize on regional and local promotions. Includes dozens of unique programs and tactics & the newest marketing ideas. **#S235...$14.95**

Welder's Handbook
Weld like a pro! Covers gas, arc, MIG, TIG, & plasma-arc welding & cutting. Weld everything from mild steel to chrome moly, aluminum, magnesium & titanium. **#S179...$17.95**

Sheet Metal Handbook
How to form aluminum, steel & stainless. Includes: equipment, bldg. interiors, layout, design & pattern-making, metal shaping, & more. **#S190...$17.95**

Stock Car Dirt Track Technology
By Steve Smith. Includes: Dirt track chassis set-up & adjustment, rear suspension systems (fifth-coil/torque arm, 4-link, leaf springs), front suspension, shocks, tires, braking, track tuning, construction & prep tips, & much more. **#S196...$24.95**

Street Stock Chassis Technology
By Steve Smith. Includes: Performance handling, chassis & roll cage fabrication, front suspension alignment, changing the roll center & camber curve, rear suspensions, springs & shocks, stagger, gearing, set-ups for asphalt & dirt, wt adjustments, track tuning, & more! **#S192...$24.95** Special Package Price for this book & Street Stock Chassis Setup Video is $58.95

Racing The IMCA Modified
Bldg the car from start to finish. How to design & build, proper suspension geometry, low bucks parts sources, rear suspensions that really hook up, complete chassis set-ups, track tuning, bldg. a claimer engine. **#S173...$19.95** Special Package Price for book & Building The IMCA Video is only $53.95

Dwarf Car Technology
By Steve Smith. Detailed chassis & set-up procedures to help you go fast! Front suspension & steering, Rear susp. & driveline, Complete chassis set-up, Track tuning. For both dirt & paved tracks. **#S225...$22.95** Special Package Price for this book & Dwarf Car Chassis Setup Video is $56.95

Chassis Set Up Videos

Chassis Setup With Tire Temperatures	#V232	$19.95
Paved Track IMCA Mod. Chassis Set-Up	#V238	$39.95
IMCA Modified Dirt Chassis Setup	#V184	$39.95
Understanding & Tuning With Shocks	#V266	$39.95
Dirt Late Model Chassis Setup	#V261	$39.95

Aerodynamics For Race Cars
Understanding aerodynamics is necessary to win. Explains: rear spoilers, downforce, front air dams, wings, drag, lift, ground effects, aerodynamics effects on handling. **#S257...$17.95**

Dirt Stock Car Fabrication & Preparation
By Joe Garrison, GRT Race Cars. Race car construction • Mounts/ brackets/interior • Wiring & plumbing • Race car body • Painting • Front suspension • Rear suspensions • Shocks • Braking system • Tires & wheels • Trans & Driveline • Chassis set-up • Track tuning • Repairing a crashed car. **#S234...$24.95**

Paved Track Stock Car Technology
Includes: Step-by-step chassis setup & alignment • Front suspension & steering • Rear susp. systems • Selecting springs & shocks • Chassis tuning with shocks • Scaling the car & adjusting the wt. • Track chassis tuning • & more. **#S239..$24.95**

How To Build & Modify Chevy S/B V8 Cyl Heads
Complete instructions for port modifications & porting techniques, creating air flow efficiency, using a flow bench, high performance valve trains, & more. Covers stock, aftermkt & aluminum heads. **#S252...$19.95**

Pony Stock/Mini Stock Racing Technology
Includes: Complete performance build-up of the Ford 2300 cc engine • Fabricating & prepping a Pinto chassis • Scaling & adjusting the wt. • Detailed chassis setup specs & procedures • Front & rear suspensions • Low buck part sources • Track tuning • For dirt & paved tracks. Covers Pro-4's too. **#S258...$24.95**

Paved Track Late Model Chassis Setup VIDEO
With Craig Raudman, Victory Circle Race Cars. Detailed step-by-step chassis setup, scaling the car, setting cross weight, alignment, choosing springs & shocks, setting bump steer, fine-tuning the chassis. Includes lots of tips & tricks. **#V279...$39.95**

The Racer's Ultimate Internet Resource Guide
The ultimate internet sourcebook for the motorsports industry. Will help you navigate the web with the most comprehensive listings of parts mfrs, suppliers, services, cybermalls and more. Includes a special section for kart racing. **#S276...$10.95**
FREE OFFER The above book (#S276) is yours FREE with every order totaling $35 or more (not including shipping and handling)!

- Shipping cost: Add $5 for first item, $8 for 2 items, $10 for 3 or more
- UPS delivery only. Provide street address
- Checks over $30 must clear first (30 days)
- Canadian personal checks not accepted
- Charge to VISA, MasterCard, AmExp, Dscvr
- In Canada add $5 to all orders
- CA residents add 7.75% sales tax
- US funds only. No COD's
- CALL FOR FREE CATALOG

Steve Smith Autosports
P.O. Box 11631- W
Santa Ana, CA 92711
Fax: 714-639-9741

- Prices subject to change without notice